T0269502

Quality management and strategic alliances in the mango supply chain from Costa Rica

Quality management and strategic alliances in the mango supply chain from Costa Rica

An interdisciplinary approach for analysing coordination, incentives and governance

Guillermo E. Zúñiga-Arias

International chains and network series – Volume 3

ISBN 978-90-8686-054-8
ISSN 1874-7663

First published, 2007

Wageningen Academic Publishers
The Netherlands, 2007

Table of contents

Acknowledgements

Time goes fast; almost five years have past and finally the work is done. This has been a period of ups and downs. At some moments I didn't notice any progress, but at others I was able to see it. This has been a dialectic process in which to go forward some time you had to go backwards and start again. Nevertheless, it has been a great time for me as a person and as well as a researcher still in development.

I would like to express my gratitude to the grant organization WOTRO, for without their support this research would not have been accomplished. I also would like to thank Wageningen University, and especially the Development Economics Group (DEC), where I was appointed and taught the academic rigor that my work should have. Of course, I am grateful for the support of all the DEC companions.

I would like to thank my promoter Arie Kuyvenhoven for his support and assistance in (re-)structuring my research. I have never seen Arie angry, but always with a smile, supporting and motivating the people at the department. I would also like to thank my co-promoter Tiny van Boekel. I met Tiny several times in Costa Rica, but for other issues than my progress in research. Tiny was always curious about how my work was going and willing to help with it.

I want to give a special thanks to Ruerd Ruben; he took a risk by selecting me, provided guidance and trust in the work we were doing. I met Ruerd in Costa Rica in 2002, during a short meeting in the Corobicí Hotel; after a small talk he told me 'see you in Wageningen in January'. And that was it. He gave me his e-mail address and a couple more for arranging the administrative part of my stay. I know he took a chance and played risk-taker, and borne potential opportunistic behaviour and moral hazard in our relationship. I want to thank him for the motivation and the hard-working process he put me in and for the trust in the work that I have been doing.

I want to express my gratitude to all the mango producers that I met during the research, I cannot mention all of them but I want to give special thanks to ASOFRUL (in Guanacaste) and to Hector Gutierrez, for their support, motivation, and collaboration in many parts of the research. In the same way, I want to acknowledge the support from Coonaprosal, ADEFA and ASOFRUPA. Without their openness and willingness to participate, this research would not have accomplished its goals.

I would like to express my gratitude to the WOTRO team, Jacques Trienekens, Aad van Tilburg, and Ruud Verkerk for pushing and motivating with comments about my work. Special thanks to Emma Kambewa and Hualiang Lu, my partners in this journey; with you two I suffered and shared the good and the not so very good parts of our research process.

I also want to thank Gert Jan Hofstede for providing the opportunity to explore gaming simulation and to Sebastiaan Meijer, who became my friend and my game board partner; I have certainly learned a lot form you.

To the MSc Students from Wageningen David Millet, Jeanina van Kampen, Maria Amparo Nieto, Sietse Sterrenburg, Rita Wells and Floor de Ruiter; many thanks for the fieldwork you conducted; your research made an important contribution to my thesis. I really enjoyed going out into the field with you, exchanging point of view and sharing our free time.

I also want to express my gratitude to the Escuela de Tecnologia de Alimentos of the Universidad de Costa Rica, for the support to the MSc students during their fieldwork, while using the laboratory tools under your supervision.

I want to thank the Universidad Nacional (UNA) and there especially CINPE, for allowing me to use the infrastructure and give me some space to do our work. Thank you very much! I want to give special thanks to my friend Fernando Saenz, although this time he was not involved in the day-to-day research. Fer, I remember that day at the Corobici Hotel when you told me 'be careful with your work, do not blow it, if you do not want to finish, I will make you finish, my word and reputation is at stake as well'. Of course I laughed but kept the idea in my mind always; you also applied to me the phrase *tenga paz* at moments when I could not take rest. Thank you very much! To Randall and Donald, for their comprehension, support, motivation, feedback, and relaxed talk. I want to express my gratitude to my friend Ricardo Fort, with whom I spent much of my time in the Netherlands, thank you for the comments on the work and especially for the company and the *trabajo en equipo*.

To my friends in Costa Rica, thanks for your motivation and support, some of you helped me to test the gaming simulations; others with the understanding of the questions of the questionnaires. And many more of you were always there to listen to me and have simple day-to-day conversations about life. Rebe, I want to thank you for the chatting and talking; you gave me a lot of time and were always there to listen and support. You have a clear idea of what studying abroad means and the effect of having to live on both sides of the ocean.

Finally, I want to thank my family in a general sense for your support. You were at my side when I needed it, not only in nice periods of life. Papi, you are an example to me in my life, both academically and as a human being. Mami, you have been always there, motivating and pushing, some times pushing more than motivating, but this is your way of teaching. José, you taught me to be persistent and to push my limits further and further. Néstor always gave me supportive words, always positive and focusing on the main goals, not the secondary targets. Ana, the youngest in the family, always asking about the things I was doing, paying attention to the small description of things, and checking for mistakes in English grammar and writing. Mary, you helped me to find peace and assisted me in getting back on track again; thank you for your company, comprehension and all the time you listened to me. This is just one step for many more to come, for you and me.

Chapter 1. Introduction

1.1 Coordination for development

Few consumers realise how many steps and procedures are required to put food on the supermarket shelf, nor do they know how it is produced, or from which towns and villages it originates. Although this is changing for some consumers who are becoming more aware of the type of food they eat, most other agents in the supply chain must meet global standards and monitor activities to be able to satisfy market demands. Supply chains for agricultural commodities have changed from supply-driven to demand-driven, implying that consumers (retailers) are the key agent in the chain. Hence, coordination, trust and governance of the chain are of vital importance to satisfy consumer preferences.

With respect to consumer preferences, there are many aspects requiring further analysis, such as prices, timely availability, presentation (i.e. ready-to-eat food) and shelf life. Many of these aspects can be addressed in terms of quality performance (understanding quality as meeting or exceeding consumer wishes). Quality is thus a complex concept that goes beyond mere consumer satisfaction. Quality has several dimensions, ranging from quality in terms of the production system (the biological and human activities for producing products with certain intrinsic and extrinsic attributes) to quality management in transport, processing and exchange activities.

Agents in the supply chain must assume the responsibility for dovetailing several activities within the supply chain in order to deliver the quality that consumers desire. For example, coordination between actors in the chain may influence the bargaining power relationships, leading to different market selection choices and changing the value-added distribution amongst agents. Otherwise, controls can be introduced in the production system to regulate products with a specific set of characteristics and standards. Product management and quality control are important for maintaining such characteristics throughout the supply chain. Therefore, quality management includes both biological management of the produce as well as human management of activities and procedures.

The mango supply chain in Costa Rica has several characteristics of interest for this research. It is a chain originally designed for the export market (where quality, management, standardisation and coordination among actors are of strategic importance). Several years have passed, and now mangoes are also very important for the local market (an outlet with lower grades and standards for quality and management, but with a high level of coordination amongst actors). The mango chain consists of several different actors: (1) small-scale producers, some of whom are stakeholders of a producers' organisation or cooperatives, and others are independent; (2) large producers who are able to produce and trade by themselves in both market segments; (3)

traders – in this category we can observe at least three actors: first, producers' organisations or cooperatives of small and medium-sized firms (owned mainly by small mango producers) whose main objective is to export mango; second, international companies, mainly multinational and medium-sized private exporters, with different degrees of integration and bargaining power toward the producers' associations and cooperatives; and third, private intermediaries that basically buy the produce directly from producers and from rejected goods coming from packing plants, and then sell the mango in the local market; (4) retailers: these actors are closest to the consumer and thus try to control the chain by exercising power over information to and from the consumer.

The present study aims to disentangle the problem related to quality management, bargaining power and coordination throughout the Costa Rican mango chain. These macro-concepts are strongly interrelated and hence difficult to unravel. As soon as one of these components is modified, the others will experience some changes as well. Essentially, the problem addressed in this research is how to improve quality performance (related to different grades and standards in sub-markets) by means of different coordination and governance regimes amongst chain agents. Due attention is given to the importance of bargaining power for controlling and monitoring the chain and its implications for the revenue distribution (that is, the derived incentives for maintaining quality over time).

It is well known that relying on only a few export commodities is a risky strategy for many developing countries. Two common problems related to this strategy are: (1) the fact that the worldwide consumption of basic commodities has grown little, while prices for the export products tend to decline (UNTAD, 2000); and (2) developing countries receive a declining share of the revenues from sales of their commodities in the global market (Morrisset, 1998; Talbot, 1997). Costa Rica began diversifying its agrarian exports some four decades ago. Any country that currently tries to diversify and enter new markets faces some common problems. First, it is difficult to increase the value of the products due to technological constraints and tariff barriers. Second, food industry processors are well established in industrialised countries, where economies of scale play an important role. Third, consumers in Western markets are increasingly looking for fresh products that are rapidly and consistently delivered through cool chains.

Under competitive market conditions, firms are faceless and nameless, demand is infinitely elastic, prices are given and the firm allocates its resources within a predetermined production function. In such competitive markets, firms maximise net profits subject to production constraints (Goldsmith and Sporleder, 1998). Unlike standard neoclassical economics, Coase recognises that there are many costs involved in using the market mechanism. These include the costs of information, negotiation and monitoring (Hobbs, 1996).

These entry barriers may easily induce market failures. To cope with the requirements of finding new market opportunities, access to information and investments in assets are of key importance. From the viewpoint of development economics, missing or imperfect markets represent a crucial step in our understanding of the economic problems of developing countries (Ray, 1998). The presence of information asymmetries, unaligned incentives between agents in the supply chain, low contract enforcement, and unequal distribution of wealth tend to increase market failures. An important information barrier caused by inadequate market institutions concerns signalling product quality and grading (Boger, 2001). Grading may result in lower information and sorting costs for products with specific quality attributes whose value is not easily measurable. Sporleder and Goldsmith (2001) explain that the signalling problem in the supply chain is bi-directional and consists of three critical dimensions, including information asymmetry, incentive asymmetry, and arduous measurability. Managerial options for signalling tend to rely on third-party protocols and procedures; differentiation through branding and reputation; insurance arrangements and warranties, and different strategic alliances and vertical integration.

It is therefore highly relevant to examine the role that governance of the supply chain plays in structuring business transactions and relationships. Willamson (1975) defined transaction costs for three main governance structures of exchange, distinguishing spot markets, vertical integration and hybrid forms. From this point of view, market transactions are arrangements between economic agents (Ouchi, 1980), but all contracts are incomplete due to bounded rationality, uncertainty, asset specificity, delivery frequency, and opportunism. Therefore, human behavioural conditions lead to market failures. Key constraints that govern the supply chain are the partition of authority and the existence of power relationships that influence how financial, material, and human resources are allocated and flow within the chain. This is what Gereffi (1994) calls the 'governance structure' of value chains. He also distinguishes between two types of supply chains: (1) product-driven and (2) buyer-driven chains. In the case of agricultural products, the latter are most common, and they are usually governed by companies that prefer vertical integration relationships controlling the marketing of the product, hence maintaining full power over the supply and delivery relationships.

Agri-food sectors of many developing countries are nowadays witnessing tendencies towards even closer vertical coordination (Costa Rica included). This implies that supply chain relationships become more closely coordinated. This occurs mainly to guarantee that technological, regulatory, and financial procedures respond to more stringent consumer preferences regarding quality and particularly safety (Hobbs and Young, 2000). Transactions and agreements are thus conducted in a defined sequence of steps from producer to final consumer.

Market segmentation is also an important device for access to premium markets (like selected export markets, fair trade, and organic foods). In order to reach different types of consumers, markets still maintain different requirements and quality parameters. Supply chain organisation involves a multitude of decisions related to marketing, economics, logistics, and organisation behaviour, which give rise to a large variety of supply chains management arrangements

(Hobbs, 1996). For understanding its effect on the structure of the industry, the causes and effects of differences in individual firm performance deserve to be analysed.

These simultaneous tendencies of supply chain coordination and product differentiation represent the background for this study. Most attention will be given to the economic, technological and managerial strategies that are used by supply chain agents to enhance agency coordination as a strategy towards more inclusive and sustainable development.

1.2 Mango production in Costa Rica

Our case study is based in Costa Rica[1], a small country with an area of 51,100 km^2, located in Central America between Nicaragua in the north and Panama in the south. Costa Rica has a population of four million people with around half of the population living in the Central Valley (composed of the four main cities). San Jose is its capital. The poverty rate is around 21.2% (18.7% urban and 24.9% rural, 2005), and the unemployment rate was 6.4% (2004). Life expectancy for 2005 was 76.1 years for males and 80.8 years for females. Social investment is 18.4% of the GDP, and the GDP per capita was US$ 4,062.4 in 2000. In the agricultural sector the GDP was 9.55% of the total (US$ 14,832.4 million, in 2000).

Historically, Costa Rica has been involved in global trade for at least 150 years, with exports of coffee and other traditional commodities. With the new settings in global trade, other products known as non-traditional exports have come into the scene. This is due to the development of new markets in response to growing consumer needs, the necessity of the government to diversify the commodity composition of trade of the country, and the possibilities of taking advantage of major geographical characteristics to access and compete in new international markets. Nowadays, non-traditional export products are most important in relation to the total exports of the country (see Table 1.1.). In 2000, non-traditional exports accounted for 48.5 percent of all exports, while in 2003, non-traditional export products already represented 56 percent of the total exports, and by 2005 they accounted for 64.2% of exports. Mango production ranged between 0.3% and 0.4% of the non-traditional export products during the past five years.

Among the non-traditional export products, we find that the export of tropical fruits is very important for Costa Rica. Pineapple, melon, watermelon, bananas, and mangoes are some of the export fruits in the export portfolio. Costa Rica has favourable natural conditions for the production of tropical fruits, and fruit exports are rapidly replacing traditional crops (such as coffee, sugar and meat) as an important source of foreign exchange. A major share of exports

[1] Sources of the data include: Banco Central de Costa Rica; Ministerio de Planificación Nacional y Política Económica; XII Informe del Estado de la Nación, Costa Rica; Carlos Pomareda, La agricultura en la economía y el desarrollo de Costa Rica, 1960-2004, en: Agricultura y desarrollo económico; Celebración de los cuarenta años de la publicación del libro Transforming Traditional Agricultura, Theodore Schultz, Eds. Grettel López and Reinaldo Herrera. Academia de Centroamérica, San José, 2005.

Table 1.1. Exports of agricultural products, Costa Rica (in millions of US$).

	2000	2001	2002	20003	20004	2005
Total exports	1697.6	1561.7	1587.8	1813.8	2024.8	2162.4
Traditional products	874.7	727.8	687.8	798.8	805.9	774.7
Non-Traditional products	822.9	833.9	900	1015	1218.9	1387.7
Mango	4.815	2.898	3.469	3.374	3.834	5.491
% Non-Traditional products	48.5	53.4	56.7	56.0	60.2	64.2
% Mango/Non-Traditional products	0.6	0.3	0.4	0.3	0.3	0.4

Source: Data collected from National Production Board (CNP), International Trade Ministry (COMEX), and International Trade Promoter (Procomer). Traditional products are coffee, bananas, sugar and meat.

is delivered to wholesalers and supermarkets in the United States, but seasonal outlets are also found in some European countries. Exports of tropical fruit represent an increasing share of agricultural production and trade from Costa Rica. Since 1992, Costa Rica has been the third biggest supplier of fresh fruit to the US market (Julian *et al.*, 2000), and fruit products represented about 35 percent of US agricultural imports from Costa Rica in 2000. Local wholesale fruit trade is cleared through the CENADA market, whereas production for local consumption is delivered at weekly farmer's markets. Fresh fruit still represents the largest share, but mango pulp is becoming increasingly important. Most processors maintain multiple marketing channels, with several exports markets (US and EU) to balance seasonal demand patterns, and a local market outlet for second-grade products. Non-traditional exports also generate considerable rural employment for the local population, both in production and processing activities. Most non-traditional export products are highly labour-intensive and require wage labour for harvesting, washing and packing activities. More than 70 percent of the labour forces are female workers (Weller, 1993).

Prospects for further increases in non-traditional agricultural export products from Costa Rica are rather promising, but also depend on the possibilities for local producers and processors to respond to higher quality and safety demands. Supermarket shoppers' concerns about pesticide residues (Ott, 1990; Huang, 1996) and nutritional attributes (Moon *et al.*, 1998) impose new restrictions on tropical fruit production, processing and handling procedures. Fresh fruit is a high-value product with a short shelf life, but it can be sold at premium prices if quality and reliability standards are met. Consumers are extremely demanding in terms of colour, size, uniformity and taste. This requires the entire channel to be well integrated through cool chains with appropriate handling and logistics, where reputation and presentation are guaranteed (Humphrey and Oetero, 2000; Kortbech-Olesen, 1997). While mango production increased between 1997 and 2006, prices have decreased steadily during the same period (see Figure 1.1.).

Diversification into non-traditional commodities involves small and medium-sized producers in the Northern and Central Pacific, partly organised in co-operatives and producers' associations. Most exports take place under contractual procedures involving specifications on maturity, appearance, size, flesh colour, skin colour, internal and external damage, bloom schedule, weather conditions, delivery frequency and chemical residues.

Mangoes have been cultivated for more than 4000 years, and were brought to Costa Rica at the beginning of the 19th century from Jamaica. The yellow varieties were the most common at that time. Export began in 1980 when the renovation of mango varieties began. Yellow mangoes were replaced by red mangoes due to changes in consumer preferences and the better resistance of red mangoes to diseases and pests. There are currently around 1995 mango producers in Costa Rica, which can be divided into three groups depending on the area of the orchards: approximately 1,170 small producers (60%) with less than 5 ha; 683 medium-sized producers (35%) with 5-20 ha, and 97 (5%) large producers with more than 20 ha. The production areas are located mainly in the Central and Northern Pacific area (departments of Orotina, Esparza, Barranca, Jicaral, Paquera, Nicoya, and Liberia).

The main quality standard that producers must satisfy is the maturity of the fruit (Peacock, 1986). This is important because in the first stage of the production process, producers are only aware of maturity; they cannot control aroma and taste on the tree and the mangoes are still green when harvested. The aroma and the flavour of the mango are monitored during

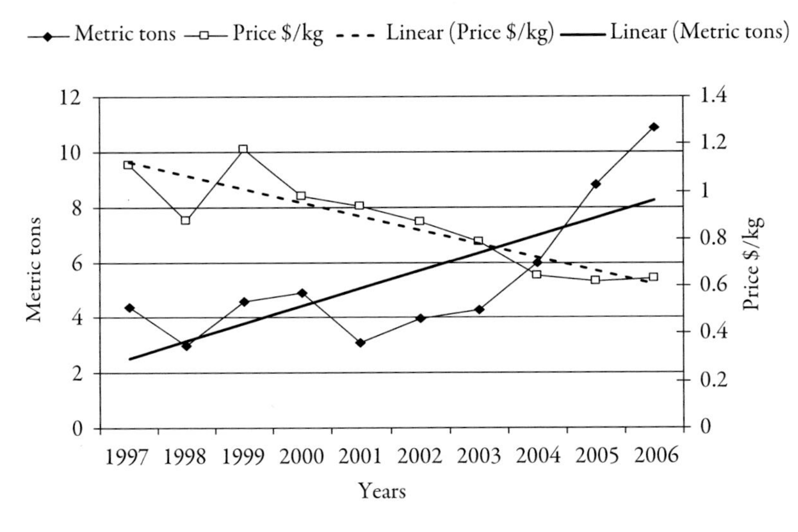

Figure 1.1. Fresh mango volume exported and prices. Costa Rica (1997-2006).

shipping to the final market (Johnson *et al.*, 1997). The indicators related to the maturity of the mango are dry matter, flesh colour, skin colour, fruit shape, Brix, bloom schedule, weather conditions, pests and diseases (Johnson *et al.*, 1997).

1.3. Research questions

The addition of new market outlets and commodity categories (in a broad sense) has been accompanied by the inclusion of new international and local regulations and transactions to handle business relationships. These new relationships tend to reinforce horizontal and vertical integration and/or supply chain coordination, based on trust, information exchange, and reputation that might reduce transaction costs and influence business performance. In this study, we aim to understand the structural and behavioural foundations for mango producers and traders to engage in particular exchange configurations relationship and the derived implications for quality performance, value added distribution and long-term competitiveness.

The main issues addressed in this thesis refer to the following research questions:
1. What are the determinants of market outlet choice for local and export market producers?
2. What kind of price and non-price attributes can be used to provide incentives for higher quality mango production?
3. How can bargaining power in the mango supply chain be assessed and what factors determine revenue distribution?
4. How does management variability affect quality variability in mangoes, and what managerial factors will contribute to quality homogenisation?

The study aims to disentangle the factors, activities and processes that have an effect on agency coordination and quality performance, by analysing interactions between physical and human activities throughout the chain. Differences in market outlets provide specific incentives for investments and management, consequently giving rise to particular configurations of bargaining power and revenue distribution between agents. Figure 1.2. shows the relationships between the different components of the research. Starting from the analytical framework of structure, conduct and performance (SCP) derived from industrial organisation theory, we focus particularly on the dimensions of quality performance (part of the so-called techno-managerial approach) and value-added distribution (derived from the global value chain perspective).

Figure 1.1. shows three different segments: (a) the value chain related to bargaining power and revenue distribution; (b) the key actors, where we have used a simplification of the original mango chain with only three actors-producers, traders/processors and consumers; and (c) the techno-managerial segment related to quality determinants and management of quality variability tailored to market selection. The first relationship refers to the interface between

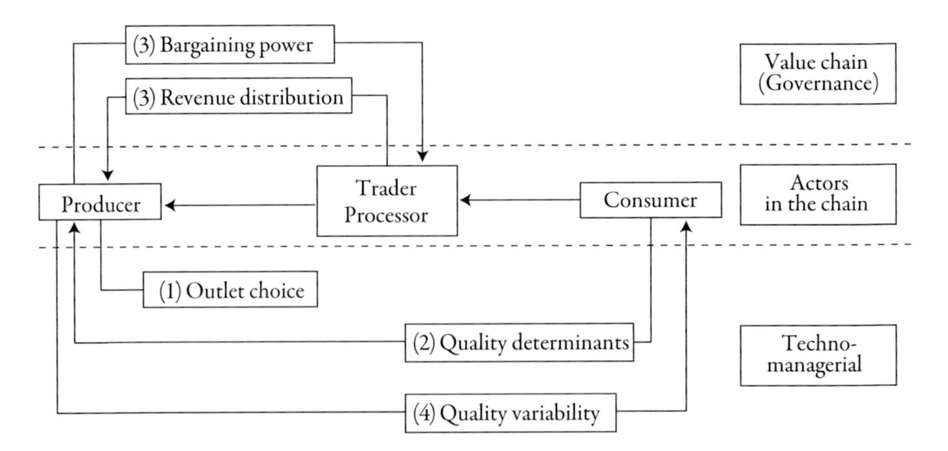

Figure 1.2. Analytical framework for the research.

producers and the traders/ processors. Producers must decide whether to produce for export or for the local market. Both markets have advantages and limitations, and producers are supposed to make rational decisions when selecting the market outlet. The second relationship refers to the interface between the consumer and the producers. Consumer preferences regarding quality must be translated into incentives and performance indicators at the production side of the supply chain and in return, producers need to provide what consumers want. This leads us to the third relationship addressed here, referred to as the determinants of bargaining power and the distribution of revenues. Bargaining power in agricultural supply chains is considered between producers and retailers, giving rise to a particular distribution of the revenues amongst different actors in the chain. The fourth relationship also has to do with the governance structure of the chain, and addresses the interface between production system management by the growers and the quality performance in the market. Quality variability is usually assumed to be lower for the export market compared to the local markets. The latter relationship is thus closely related to the selection of market outlets and different consumer wishes concerning quality in specific market segments. Whereas we analyse each of the relationships in a partial manner, within the supply chain they are closely interlinked.

1.4. Research methods

Our study is based on the integration of different analytical approaches to address quality, supply chain management and marketing issues simultaneously. Quality of produce is of great importance for consumers, but quality is not a simple concept that we use to indicate whether something is good; it is more than that. Quality is composed of internal and external attributes of a certain product, as well as a relative concept related to consumer preferences and wishes. Quality can be constructed as a continuous process of improvement where consumer preferences, commodity characteristics and technological options are integrated into one system. Quality function deployment (QFD) provides an analytical procedure that helps to translate customer demands into the appropriate technical requirements for each stage of product development and delivery (Sullivan, 1986). Luning *et al.* (2002) stress – in relation to business performance – that quality cannot merely be considered as physical quality (intrinsic and extrinsic attributes), but must also include other dimensions of competition (e.g. cost, availability, flexibility, reliability, total service). Therefore, quality is a mixture of biological activities and human activities, both related to management organisation, referred to as the techno-managerial approach.

It is possible to combine these approaches within an industrial organisational paradigm. Bain (1951) developed an analytical framework called structure, conduct and performance (SCP). This is a framework used to study an industry, and it states that the market structure (i.e. degree of concentration or governance in the market) will induce a reaction from the participants – conduct (i.e. barriers of entry or exit, advertising and pricing) – and this conduct will subsequently affect the performance of that industry (i.e. market power, profits, and revenue distribution).

The three dimensions are clearly interrelated: SCP emphasises that the structures of the market and the environment in which the firm is involved will cause a reaction, in terms of behaviour, to the structure and this conduct will thus affect the firm's performance. QFD refers to the activities to develop high(er) quality products and outlines performance indicators in the production system. Taking into account the consumer preferences, the firm will react and will rely on the available market structure to produce what the customers are asking for. It is clear that coordination is needed to be able to align the incentives between the different actors in the supply chain.

The global commodity chain (GCC) deals with this coordination problem taking consumer preferences as the starting point for most agricultural (consumer-driven) chains and this approach also reveals the distribution of power in the chain, the value added and market selection problems. The more generic approach of supply chain management (SCM) addresses the relationship between transactions and human behaviour, related to information, product and financial flows throughout the chain. SCM offers an opportunity to capture the synergy of intra- and inter-company integration and management, thus dealing with the total activities of the business (Lambert *et al.*, 2000).

To guarantee adequate dovetailing of the afore-mentioned analytical approaches, we used SCP in a slightly different way. First, we analyse the SCP not of an industry, but more at the micro-level of transactions. Second, we assume that mango supply chain agents in Costa Rica operate in a competitive market. However, using institutional economics we posit that markets are not frictionless and consequently market failures occur due to transaction costs. We thus apply transaction cost economics, considering the co-existence of different forms of organisation or governance. The frictions inside the market are the transaction costs that are basically related to transactions themselves. By accepting the existence of transaction costs, certain types of conduct will induce or cause governance structures to operate. Third, we use additional performance indicators for conduct; in addition to the usual including income and profitability measures, in this study other performance indicators will be related to quality measurement, bargaining power, and revenue distribution.

The data collection and analysis is based on three methodological approaches: (1) survey analysis, (2) gaming simulation analysis, and (3) laboratory experiments. Surveys and laboratory analysis are applied in the analyses related to market outlet choice, and to identify economic incentives for quality management and for dealing with variability in quality. For the analysis of bargaining power we used the gaming simulation approach.

We conducted three field surveys, deriving both socio-economic information, production process characteristics, and agency perceptions about the mango business in a comprehensive way. The first survey covered nationwide indicators of mango production. A second survey focused at the farm level with mango producers and a third survey has been carried out at the level of supply chain relationships with producers, processors and retailers in the sample. We have treated data quantitatively, and conducted several statistical analyses, such as regression analysis and ANOVA analysis, in addition to the common descriptive analysis of the data set.

The other methodological approach is based on simulation gaming. In the past, this technique has been mainly used in learning processes; our innovation is to apply it for research purposes as well. We developed a generic supply chain board game to be played with the real mango producers and with this we analysed the perceptions on bargaining power and the outcomes in terms of revenue distribution. One of the main advantages of this tool is that people are invited to behave like they do in real life but in a controlled environment. The outcomes of the game simulation sessions have been recorded using pictures, questionnaires, contracts, interview forms and debriefings during and at the end of each session.

We also collected mango samples from the producers and from other actors in the chain and we made a laboratory analysis of the real intrinsic and extrinsic quality attributes of the mangoes. With these quality attributes, we were able to develop a quality index. These quality attributes of mango could be related again to the socio-economic factors of the surveys. In addition, we also conducted a consumer panel discussion to find out consumer preferences regarding mangoes (only for local consumers). We analysed the data using several multivariable techniques, such as principal component analysis, regression analysis, ANOVA, discriminant analysis and categorical regression, Tobit regression and logistic regression, as well as the common descriptive analysis of the data set.

1.5. Outline of the thesis

The study is structured into four main chapters. We start with an analysis of the structure of the mango supply chain in Costa Rica, presented in chapter 2. An explanation is given about reasons for market outlet choice, as a basis for understanding the differences between export and local market orientation. Hereafter, the decision-making process and governance of the chain are discussed. Different actors are simultaneously involved in different mango outlets. Producers thus face a strategic choice between (a) market outlets devoted to exports where quality attributes such as size, sugar content, absence of external and internal damage are key determinants for the transaction and the business relationship; and (b) local markets (composed of wholesaler, retailer, and middlemen, among others), where different qualities and delivery modes can be accommodated. We argue that market selection depends on farm-household characteristics, the production system, price attributes and the market context. The selection of a certain type of outlet is also related to specific contract configurations (i.e. quality control, payment mode, type of agreement, volume, rejection rate, etc.). To this effect we disentangle factors that influence market differentiation between the export and local market in the Costa Rican mango sector; and we identify major determinants for the differences in outlet choice.

In chapter 3, we analyse the economic incentives for improving mango quality. They include price and non-price attributes, as well as household and production system characteristics. Although we focus the analysis on the producers side (i.e. laboratory measurement of mango quality and field survey of the characteristics of producers), for this part of the study we will also take into account consumer preferences. This study is based on an integrated methodology for identifying effective economic incentives to enhance quality performance by mango producers in Costa Rica. We analyse the relationship between intrinsic quality characteristics measured at the field level and the socio-economic characteristics of the mango producers in the Central Pacific zone of Costa Rica. A mango quality index for specified export market outlet is constructed and subsequently related to a set of socio-economic characteristics at farm and household level. We identify the relationships between farm-household characteristics, production system features, marketing relationships and quality attributes measured by consumers' quality perceptions and preferences. The general attributes of the quality index – mainly ripeness, appearance and variability – are supposed to be related to farm-household characteristics such as producers' age and experience, input use intensity and family labour availability. In addition, preferences for certain contractual regimes and marketing arrangements may give rise to differentiation in quality performance.

Chapter 4 deals with the factors that affect bargaining power of the agents involved in the chain and its effect on the revenue distribution throughout the chain. We thus analyse the implications and effects of the market structure for the distribution of power and revenues. Transactions between producers, traders, retailers and consumers in the mango supply chain are governed by contractual arrangements concerning outlet choice, price, volume, quality, and delivery frequency. We designed a gaming simulation framework for the appraisal of different sets of delivery transactions between mango supply chain agents. The gaming simulation design closely mimics the negotiation conditions in the Costa Rican mango supply chain. Agency roles have been defined for all participants in the supply chain and attributes of all transactions will be recorded. This allows for the identification of the production and delivery choices made by agents under different market configuration conditions.

In chapter 5, an analysis is presented of the relationship between the variability in quality and the variability in management practices for main actors involved in mango trade. We assumed that management activities have a positive effect on the reduction of the variability in quality. Quality is a key aspect for evaluating the performance of commodity chains, but quality management is a highly complex procedure. Quality refers to the consumer perception of a certain product, as well as the objective relationship between intrinsic and extrinsic attributes of that product. Improved cultivation and general management activities may enhance the quality and/or reduce the variability of the produce. In addition to these technological measures, timely access to information and improved quality management operations can be helpful to reduce human-related variability. We therefore provide an explorative framework for disentangling the interactions between different managerial activities that affect the variability of the quality. For this purpose, we used several data dispersion statistics to understand the impact of technological and socio-economic factors on the heterogeneity in quality performance at different stages of the supply chain. We conducted a field survey amongst 51 stakeholders involved in the mango supply chain in Costa Rica and collected information regarding their production technologies, agro-ecological conditions, management intensity, quality control, contracting practices and marketing operations. We also collected 10 mangoes randomly from each actor to analyse the variability in intrinsic quality attributes (defined as the ratio between Brix and pH). We argue that quality variability is influenced by both technological and socio-economic variation. In the mango supply chain from Costa Rica, the management differences amongst actors vary depending on their closeness to the consumer market. Actors closer to the consumer tend to maintain higher variability in their management practices in order to be able to respond better to market challenges.

Finally, chapter 6 provides a synthesis of the findings derived from each of the specific analyses and indicates some promising pathways for improving supply chain coordination and quality performance in the mango sector of Costa Rica. Most attention is given to opportunities for dovetailing strategic outlet choice, supply chain management and bargaining procedures that permit simultaneous improvements in product quality and value-added distribution, in such a way that incentives are aligned between market opportunities and agency behaviour.

Chapter 2. Determinants of market outlet choice of mango producers in Costa Rica[2]

2.1 Introduction

The organisation of the mango supply chain in Costa Rica includes a large number of different agents that are simultaneously involved in numerous transactions and oriented towards multiple market outlets. Mango transactions differ in terms of volume, quality, price and delivery frequency, while the fruit can be sold both at the local open market and through wholesalers, as well as at international markets through multinational trading companies. Relationships between producers' associations, (local and international) traders, retailers and consumers are structured through a complex sequence of delivery transactions.

Mango (*Mangifera indica* L.) is an exotic tropical fruit in high demand around the world. In Costa Rica, it was originally used for home consumption as well as to provide shade in plantations and homesteads. In the early 1980s, mango was increasingly considered by smallholders as an opportunity to grow a commercial product for the export market.

There are nowadays about 150 different varieties of mango known worldwide, but only a few varieties are of commercial importance (i.e. Haden, Irwin, Keith, Tommy Atkins). The first imports of mango to Costa Rica took place in 1796 from different Caribbean countries (Lezema, 1989). In the early 1970s, the University of Costa Rica (UCR) introduced some special red and yellow varieties in the Orotina region for commercial purposes (Mora *et al.*, 2002). At the beginning of the 1980s, the production of several varieties of red mangoes for (European) export markets became more important, thus leading to a local reduction of the area devoted to yellow mangoes (Jirón, 1995; Buzano, 1997).

The cultivated mango area in Costa Rica has remained stable for some years. The Ministry of Agriculture (SEPSA, 2001 and CNP, 2006) estimates some 8,200 hectares as the total area of land used for mango production. Total production is approximately 32,000 metric tons, of which 34 percent is exported. Costa Rican fresh mango exports are mainly directed to Europe (86.3%) and to the USA and Canada (13.3%). The remainder of the mangoes (both fresh and processed) are consumed in the local market.

The Costa Rican market configuration is characterised by a simple product flow for the export market, but includes far more complex configurations for sales in the local market. For the

[2] An earlier version of this paper was published as: G. Zuniga and R. Ruben, 'Governance regimes for quality management', published in: R. Ruben, M. van Boekel, A. van Tilburg and J. Trienekens (eds) (2007) Tropical Food Chains, Wageningen, Wageningen Academic Publishers, The Netherlands.

export market, most of the fruit originates from producers' associations that deliver to a cooperative, which packs the mangoes and sells them to the exporter. In the local market, there are many independent producers, as well as organised producers and intermediaries. Some of them deliver fruit rejected by the export market to several local outlets, including wholesalers, local markets and also consumers.[3] Local market outlets include deliveries to the wholesale market (called *Cenada*) and sales at the open-air farmers' markets. Whereas some producers are exclusively oriented to either the export or the local market, a significant number of farmers deliver to both market outlets, thus taking advantage of (seasonal) quality and price differences.

The basic determinants of the buyers' selection and outlet choice are related to price and other attributes. The export market is a specialised market where quality, price and geographic location play an important role. The wholesale market is characterised by more flexibility in product administration and quality, where normally there are many buyers and sellers, and producers can easily meet other producers or an intermediary. Prices are mainly determined by demand, sometimes by previous agreements and also depend on the quality characteristics of mangoes.

The local open-air farmers' market is a similar outlet, but producers can usually get higher prices because they sell small amounts and maintain contact with many different buyers. For smallholder producers this appears to be an attractive outlet, since many of them are sharecroppers and therefore have small amounts of different products to sell.

Intermediaries are key actors in the process of local market differentiation. They buy the fruit from farmers and deliver it to the Cenada wholesale market or to the farmers' market. In addition, they can buy rejected mangoes and surplus from producers oriented toward export markets and sell them to retailers. Intermediaries thus possess much critical information about the mango sector and control transport facilities as well, since they have their own trucks.

McLeay and Zwart (1998) indicated that for an individual farmer sales transaction choice is influenced by marketing competencies and strategy, farm and farm manager characteristics, and the structural characteristics of the industry in which the transaction is taking place. Market outlet choice processes include three broad categories: external, internal and mixed (Koch, 2001). In the external category, trade barriers are of great significance (i.e. international regulation for exports). Strategic objectives and experience are important dimensions in the internal category, and some of the dimensions in the mixed category are resource access, proximity to the market and the market portfolio. Along the same lines, Ellis and Pecotich (2001) summarise the international process as a social factor as well, where exports cannot be initiated without the coexistence of three conditions: (1) the capability to go abroad, (2) the motives to go abroad, and (3) the awareness of a particular market opportunity.

[3] The main quality criteria for rejection are related to external damage (appearance, black spots, bruises) and internal damage (fly ridden fruit, rotten seed, among others), or due to the misuse of pesticides.

In this chapter, we address the external, internal and mixed categories affecting market choice at the producer level. One of the characteristics of the Costa Rican mango market is that exporter-oriented producers can deliver to both markets, while local market producers can only deliver to the local market. In our analysis we follow the same line of thought and explore the factors that influence market differentiation between the export and local market in the mango sector of Costa Rica, identifying the major determinants for differences in outlet choice. We first discuss the importance of price attributes, production systems organisation, farm-household characteristics and market context for outlet choice decisions. Hereafter, we present our empirical survey of outlet choice decisions amongst a sample of 94 farmers in the largest mango-producing region (Central Pacific zone) of Costa Rica. We use a Tobit model and ANOVA analysis for the empirical analysis of the determinants of market outlet choice. Finally, we discuss the structural, institutional and behavioural factors that typically determine farmers' choice for a specific market channel orientation.

2.2 Factors influencing market outlet choice

Market outlet selection is a key task for everyone in the supply chain. Agents must find business partners that meet the minimum requirements of the market and the firm. The primary rationale for market segmentation is to identify the segments that are most interested in specific commodities and focus marketing efforts on them in the most effective way (Jang *et al.*, 2002). Kotler (1999) defines market segmentation as the subdivision of a market into different subsets of customers, where any subset may conceivably be selected as a target market to be reached with a distinct marketing mix. Profitability and risk, as well as variability and accessibility thus play a vital role in evaluating the attractiveness of each segment and selecting the best target market (McQueen and Miller, 1985; Jang *et al.*, 2002).

Traditional economic theories on comparative advantages have contributed significantly to an understanding of trade at (inter)national levels, but do not fully illuminate the forces driving major business between similar countries in the same industry (Brewer, 2001; Dunning, 1988). This incompleteness of the classical explanation of trade led to the behavioural school (Kay, 1993; Porter, 1990, among others), which asserts that within economic parameters, it is the judgement and decisions of firm managers that defines internationalisation and its consequences (Chetty and Holm, 2000 quoted by Brewer, 2001). It has been argued that firms are generally not entirely rational in international market selection and that market outlet choice is a very unpredictable, disjointed process (Toornoos, 1991). However, producers tend to rationalise the choice on the basis of objective information and by comparing return rates. According to Granovertter (1985), this apparent irrationality is due to the ignored social context in which economic exchange is embedded.

From the evolutionary point of view, it is not the utility level that matters in market selection, but the chances of survival (Amir *et al.*, 2005). This evolutionary principle leads to the consideration of the process of natural economic selection among participants, or among the market participants, or among the strategies of behaviour they adopt (Alchain, 1950; Enke, 1951; Penrose 1952; quoted by Amir *et al.*, 2005). Morgan and Katsikeas (1997) have identified four sets of obstacles that explain why domestic firms are discouraged from exporting. The first is strategic and is related to a firm's ability and capability to look for a good, reliable partner. Second, operational and logistic obstacles can be seen in small firms that are not able to export directly to the general wholesaler market due to economies of scale. In addition, according to the perishability of the commodity they produce, specialised logistic methods and procedures must be used. Third, other obstacles are related to the limited access that smallholders have to information, and therefore they generally depend on the buyer for information about the market. Fourth, process-based obstacles such as innovation, technology and the use of international standards are critical for small producers willing to export. These technologies, innovations and standards are usually expensive and hence prohibitive for smallholders.

From the institutional economics point of view, Fafchamps (2004) explains that product exchange can be organised in three different ways: via gifts (intra-household, with families and sometimes friends), with a market-oriented approach (based on reciprocity and pursuing self-interest) and by means of hierarchies (firms and government; relying on command and control). We have developed an analytical framework to explain the interactions between the institutional structure where market decisions are taken and the behavioural outcomes of market selection (see Figure 2.1.).

In the following, we will discuss the main components of this framework systematically and identify the appropriate operational definitions for each of the variables.

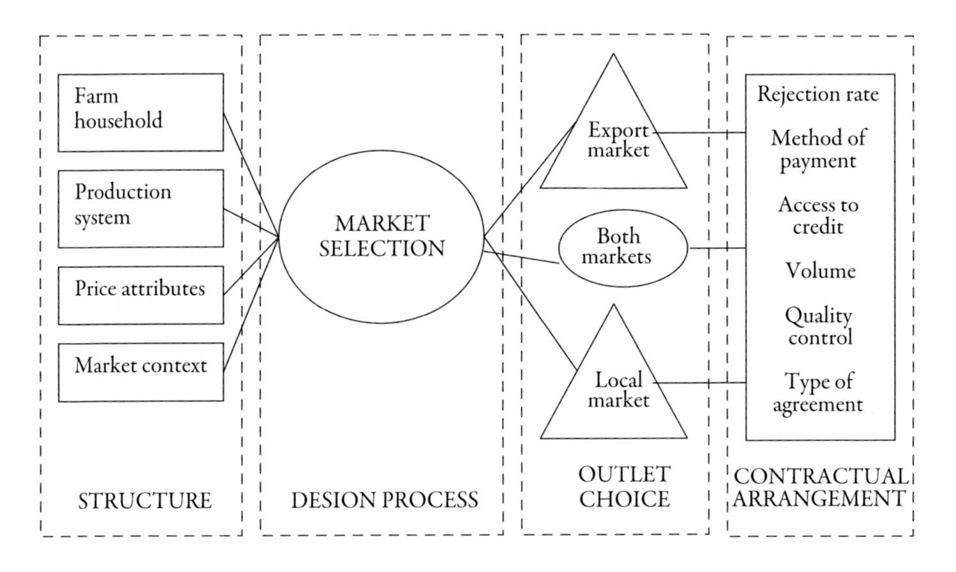

Figure 2.1. Analytical framework for market outlet choice.

2.2.1 Farm-household characteristics

Market outlet choice is likely to be influenced by farmers' attitudes to risk (Agarwal *et al.*, 1992). When actors face uncertainty, they will turn to others whom they know and trust (Galaskiewicz, 1985; Podolny, 1994). Baker (1994, quoted by Tenbrunsel *et al.*, 1999) asserts that relationships are a fundamental human need, whereas Coleman (1988) argues that social context shapes, redirects and constrains a person's actions; consequently, personal relations (shaped in a specific social context) will reduce chances for opportunistic behaviour and moral hazards (Baker, 1990 and Ben-Porath, 1980) and are likely to enhance cooperation (Granovetter, 1985). In the case of Costa Rican mango production, farmers select market outlets depending on their social network and background, and hence seek to reduce transaction costs and uncertainty, but they feel safer in a market setting where they maintain long-term relationships with participating agents. In addition, Sáenz-Segura (2006) found that the location of the orchards (proximity to the customers or buyers) defines the market outlet choice of the producer.

2.2.2 Production system

Agarwal and Ramaswami (1992) emphasise that potential sales (i.e. market size and growth rate) is an important business factor influencing market selection. Sáenz-Segura (2006) and Sáenz and Ruben (2004) find that scale of production and experience are positively related to the choice of export markets. Location-dependent costs arise from raw material acquisition and heterogeneous costs of operation at different sites (Rhim *et al.*, 2003). Ohyama *et al.* (2004) stress that an open economy regime creates a mechanism that selects new entrants in the order of their entrepreneurial abilities. This implies that entrepreneurs that are more capable run larger firms, and it is expected that early entering firms have a larger scale of production than later entrants under the open economy regime. Regardless of their type, older (mature) firms tend to be larger than new firms (Agarwal and Gort, 2002).

2.2.3 Price attributes

Price is usually considered as one of the most important attributes in the (neo-classical) analysis of economic regimes; according to Williamson (2002), the chief mission of neoclassical economics is to understand how the price system coordinates the use of resources. Sáenz-Segura (2006) and Sáenz and Ruben (2004) find in Costa Rica that prices for products in the export market are higher than those in the national market. Most producers, including mango producers, allocate part of their production to the domestic market for different reasons: first, to ensure an alternative source of income (flexibility), second to gain bargaining power (reducing hold-ups), and third to reduce the losses in production (decrease risk and uncertainties, in addition to being able to sell the fruit rejected by the export market). As asserted by Wilson (1986), farmers rely on market diversification as a protective device. Whereas prices remain important for outlet choice, they are certainly not the only device and strategic factors may lead to a marketing mix based on different outlets.

2.2.4 Market context

Three main factors for market selection have been described by Brewer (2001), namely business factors, chance of success and physical distance. *Business factors*: many firms may not engage in exports either because they do not have the necessary resources or because they do not want to commit themselves (Christensen, 1991). Resource scarcity can restrict the ability of small firms to enter the export market (Moen, 1999). Papadopoulos *et al.* (2002) identify demand potential and trade barriers as major determinants for the market selection strategy followed by the firm. Demand potential is related to market similarity and consumer behaviour, and differentiates between substitute and complementary commodities. Trade barriers, both tariff and non-tariff barriers, and entrance barriers, together with geographical distance have implications for transaction costs. Hart and Holstrom (1987) and Chiarelli *et al.* (2002) state, however, that other purchasing conditions such as terms of payment, provision of credit for inputs, frequency of delivery, seed, and technical assistance could enable producers to improve their product quality and thus influence their options for alternative market outlet choice. Sáenz-Segura (2006) and Sáenz and Ruben (2004) found indeed that in Costa Rica the availability of credit tends to increase the export orientation.

The *chance of success* factor has been explained by Erramilli and Rao (1990) and Terpstra and Yu (1988) from the fact that firms try to follow competitors as well as clients into new markets. Hoang (1998) finds that small firms are usually more reactive than pro-active in market selection issues. Regarding *physical distance*, Driscoll and Paliwoda (1997) found that the socio-cultural distance between economic agents is important for entry mode choice without consideration of experience. Other authors such as Andersen and Buvik (2002) and Papadopolous and Denis (1988) explain, though, that physical distance often results in targeting markets close to the firm's immediate neighbours, since geographic proximity is likely to imply more knowledge about markets and guarantees easier access to information.

To summarise our understanding of the factors that influence market outlet choice as derived from the literature review, in Table 2.1 we present an overview of some of the major attributes that were found to be significant factors for the internationalisation of activities in selected field studies. These variables will also be considered in our subsequent data analysis. We follow the analytical framework presented in Figure 2.1. by disaggregating the structural characteristics of the producer, the production system, market context and the price regime that will affect the producer's outlet choice (see Figure 2.1.).

Table 2.1. Variables and dimensions influencing market outlet choice.

Dimension	**Structural characteristics**	**Operational variables**	**Literature (sources)**
Price attributes	Price	Average price (in colones)	Saenz-Segura and Ruben (2004); Saenz-Segura (2006)
Production system	Size	Orchard Size (in ha)	Agarwal and Ramasnani (1992); Agarwal and Gort (2002)
	Scale	Mango production in boxes	Saenz-Segura and Ruben (2004); Saenz-Segura (2006)
Farm-household characteristics	Risk	Producers' risk attitude (1-10)	Agarwal and Ramasnani (1992)
	Experience	Years of producing mango	Saenz-Segura and Ruben (2004); Saenz-Segura (2006)
	Trust	Trust in buyers (1-10)	Galaskiewicz (1985); Podolny (1994)
Market context	Distance	Distance to market, geographic location	Brewer (2001); Driscoll and Paliwoda (1997)
	Purchase condition	Type of agreement, (1 none, 2 verbal, 3 written agreement)	Papadopoulos *et al.* (2002); Chiarelli *et al.* (2002)

2.3 Materials and methods

The empirical material for the analysis of market outlet choice is derived from a representative survey amongst mango producers in the major production region of Costa Rica. The total sample included 94 farmers selected randomly from the producers' census for the Central Pacific zone (CPZ) of Costa Rica, carried out by the National Production Bureau (CNP) in 2003 (see Box 2.1.).

Data collection was based on a questionnaire given to farmers regarding farm-household characteristics (i.e. family size, education, producer's age and experience, cultivated mango area) and production system characteristics (i.e. soil fertility, technical training, road conditions, conditions of mango farms, whether there were any trees stimulated by chemicals and physical treatment). It was followed by a series of interviews with producers to reveal their preferences regarding outlet choice and relationships with traders. We also collected detailed data on the type of delivery agreements (verbal or written), payment procedures (cash, cheque), rejection rates and the standing arrangements between buyers and sellers (i.e. visits to the farms for inspection, quality control in the orchards, credit delivery, among others).

Box 2.1.

The census included data from 1500 producers. After cleaning up the data and grouping the producers by area, the valid number of observations was reduced to 861 producers. We used the census from 2003 to determine the appropriate sample size. To calculate the sample size the following formula was used:

$$n_o = [(Z_{\alpha/2}*\sigma) / d]^2 / \{1 + (1/N) * [(Z_{\alpha/2}*\sigma) / d]^2 - 1]\}$$

where $Z_{\alpha/2}$ = Confidence Interval of 95% (1.96); σ = Standard Deviation of Population (10.86); d = permitted error (2); and N= Population Size (861). After calculation, the appropriate sample size resulted in 101 producers. Hereafter, a correction formula for finite populations was applied:

$$n = n_o / [1 + (n_o / N)]$$

The required sample size is 91 mango producers. Stratified random selection of producers was applied with the assistance of extension agents of Agriculture and Livestock Ministry (MAG) and the CNP. Within the Central Pacific sub-regions, five areas were selected to conduct the survey: Orotina, San Mateo, Esparza, Garabito and Puntarenas (Central). Stratified sampling methods were used to define the number of sites and producers by location. The total number of valid mango producer interviews was 94.

Trust between producers and buyers was assessed on a Likert scale from 1 to 10, as were risk attitudes. Finally, some key market outlet characteristics (such as the number of buyers, the years already attending the market, distance to market, price information and the degree of trust between seller and buyer) were included in the survey.

To understand the main differences amongst agents delivering to certain market outlets, an ANOVA analysis was performed. This type of analysis was used because we wanted to identify the key differences between the economic agents that deliver to different outlets. For this purpose, ANOVA exhibits some advantage for analysing situations in which there are several independent variables, revealing the way that these independent variables interact with one another and the effects of these interactions on multiple market outlet choice as a dependent variable (Field, 2002). We relied on a Gabriel's post hoc procedure to check for significant differences among the different outlets, while considering different group size in the independent variable.[4]

To disentangle the attributes that differentiate market outlet choices, a Tobit analysis was carried out. Based on the review of the literature, we selected a number of fixed producer

[4] A post hoc test consists of pair wise comparisons of all the different combinations of treatment groups (Field, 2002).

characteristics for the Tobit analysis. As an independent variable, we used the percentage of production delivered to the export market in a range from 0 to 100. The Tobit model resembles the Probit analysis, but the dependent variable is continuous, not binominal (0 or 1). A Tobit analysis is now easy to perform due to its incorporation in several statistical packages, resembling in many features a linear regression procedure (Greene, 2003). We used STATA software for the data processing and analysis.

2.4 Results

The data analysis is divided into three parts. First, we conducted an analysis at the level of the different agents (mango producers oriented to the export market, to the local market or to both markets) to understand the major intrinsic characteristics of farmers operating in each of the market segments. Second, we analysed the structural factors that explain differences in outlet choices, using the results of the Tobit analysis. Third, we have provided a description of the factors that underpin the strategic interfaces between supply chain agents in the mango chain, paying particular attention to the Costa Rican chain configuration. The latter analysis is based on ANOVA tests.

2.4.1 Producers' characteristics and outlet choice

In order to characterise the different types of mango producers, we can divide the sample population into three groups: export market-oriented (EM), local market-oriented (LM) and agents participating in both markets (BM). Table 2.2 present the results of the analysis of significant intrinsic differences between producers operating within each market segment.

There are few differences between farmers in terms of their farm-household characteristics (see Table 2.2). The producer's age, size of the farm, and family size have no significant impact on outlet choice. More mango experience leads to preferences towards the local market, whereas farmers oriented towards export markets tend to possess less experience with mango production. This counter-intuitive result may be explained in part because mango exports are rather recent in Costa Rica and are mainly promoted at newly established plantations. The level of education does have a positive relationship to export market orientation.

Regarding production system characteristics, it is important to note that export-oriented producers perceive themselves as operating under more favourable weather conditions than those of producers delivering to the local market. Moreover, the former tend to follow the advice from agricultural outreach workers more carefully, given the fact that strict maintenance of technical production procedures and general agricultural practices (such as Eurep-GAP) are of key importance for sustained access to export markets and retail outlets.

Price characteristics appear not to be significantly different for market choice orientation. Producers' input costs and output prices are somewhat similar in various market channels,

Table 2.2. Producer characteristics underlying market outlet choice

Dimension	Variable	Difference EM - LM	Difference EM - BM	Difference BM - LM
Farm-household characteristics				
	Producer's age			
	Producer's experience	*(-)		
	Mango area			
	Family size			
	Producers' education	**(+)		
Production system characteristics				
	Weather conditions	*(+)		
	Farm fertility			
	Farm conditions			
	Tree stimulation			
	Follow technical advice	*(+)		
	Mango training			
	Road conditions			
Price characteristics				
	Inputs cost			
	Mango price			
Agreement characteristics				
	Rejection rate	**(+)	***(+)	
	Type of agreement	***(+)	***(+)	
	Method of payment	***(+)		***(+)
	Seller visits to buyer			
	Buyer visits to seller	***(+)	***(+)	***(+)
	Buyer's supervision in orchard	***(+)	***(+)	
	Quality control in the orchard	***(+)	***(+)	
	Access to credit	***(+)		
Market characteristics				
	Number of buyers	*(-)		*(-)
	Years attending the market	***(-)		**(-)
	Trust towards buyer			
	Distance to market	**(-)	***(-)	
	Mango dependency			
	Knowledge on mango price			

Table 2.2. Continued

Dimension	Variable	Difference EM - LM	Difference EM - BM	Difference BM - LM
Agreement characteristics				
	Rejection rate	**(+)	***(+)	
	Type of agreement	***(+)	***(+)	
	Method of payment	***(+)		***(+)
	Seller visits to buyer			
	Buyer visits to seller	***(+)	***(+)	***(+)
	Buyer's supervision in orchard	***(+)	***(+)	
	Quality control in the orchard	***(+)	***(+)	
	Access to credit	***(+)		

Significance at *** 1%, ** 5%, , * 10%
+ = positive relationship; - = negative relationship

and thus major differences in delivery conditions are expected to be determined by non-price attributes, related to the market configuration and the characteristics of contractual arrangements.

With respect to the agreement characteristics, we find that access to the export market requires the producer to comply with higher standards. Quality surveillance from the buyer in the orchard is one of the main differences among market outlets. Moreover, in terms of the associated contract characteristics for different outlets, written contracts are more frequent in the export market, while cash payments are more common for the local market. In addition, producers delivering exclusively to export markets have better access to credit. The main differentiation in market outlet choice thus appears to be related to the more demanding delivery conditions in the export market compared to the local market, which result in differences in production and marketing behaviour.

Although it might be reasonable to expect that mango dependency will be higher for producers specialising in export market production, this appears not to be true; there are no significant differences between outlet orientation and mango income dependency. Local market-oriented farmers can deal with more buyers and have usually attended the market for a longer period, thus being able to offer more bargaining options. An important feature for export market-oriented producers is that they are located closer to the market. For integrated mango supply chains, the nearby location of export processing facilities near the orchard provides incentives for prompt delivery and prevents free-riding. Nevertheless, trust relationships with the buyers do not present any significant difference between the market channels, or with regard to knowledge about the mango price.

2.4.2 Mango export outlet choice selection

We can proceed with a more detailed analysis of the determinants of mango outlet choice, addressing the factors that influence deliveries to specific market segments. We have used a Tobit regression model with the percentage of delivery to the export market as the dependent variable (see Table 2.3). Results show that the significant explanatory variables include several *farm-household* characteristics (mango experience and risk attitude), some aspects of the *production system* (mango area, scale), and *market context conditions* (purchase conditions, geographical location and distance to market).

We can consider these results against the background of the structure of the mango market in Costa Rica (Section 2.1.), taking into consideration the theory available on factors influencing market outlet choice (see Section 2.2.).

Table 2.3. Structural factors determining export orientation (Tobit model).

	Coefficient	Std. error	T statistics	Prob.	Sign.
Farm household					
Mango experience (years)	-5.012	2.460	-2.04	0.044	**
Risk attitude (Likert 1 to 10)	-120.620	66.036	-1.83	0.071	*
Trust (Likert 1 to 10)	-11.870	8.999	-1.32	0.191	
Price attribute					
Average price (colon/box)	-0.009	0.020	-0.45	0.653	
Production System					
Mango area (ha)	4.395	1.935	2.27	0.026	**
Scale of production (boxes/year)	0.001	0.001	2.22	0.029	**
Market Context					
Purchase condition (1 no agreement, 2 verbal agreement, 3 written agreement)	104.743	34.967	3.00	0.004	***
Geographical location (0 Alajuela, 1 Puntarenas)	140.047	53.355	2.62	0.010	***
Distance to market (km)	-1.008	0.486	-2.07	0.041	**
Constant	-0.076	131.827	0.00	1.000	

Significance at *** 1%, ** 5%, * 10%.

Adj. $R^2 = 0.277$, Number of obs = 94, LR $Chi^2 = 53.05$, Prob > $Chi^2 = 0.00001$, Pseudo $R^2 = 0.1846$.

2.4.3 Farm-household characteristics

Producers eligible for mango deliveries to the export market are usually strictly selected. We therefore expect from the literature that previous mango experience, risk-acceptance attitude and trust with the buyers would contribute to more export-orientation. Our results are, however, not fully consistent with the literature. First, in our data, trust appears not to be significant. Galaskiewcz (1985) and Podolny (1994) already emphasised that trust in business partners is a key determinant for the internationalisation of supply chains. In the case of mangoes from Costa Rica, trust may be of less importance for market outlet selection, since most agents involved in the transactions have met one another only recently and business relationships are still limited to short-run deliveries with no clear perspective for long-run arrangements. In addition, there is a clear lack of information regarding the reputation of agents, and this causes the uncertainty in transactions to be relatively high; agents are more likely to search for arrangements that safeguard the delivery and may rely on interlinking transactions (i.e. input credit and technical assistance) to prevent opportunistic behaviour and moral hazards.

With respect to risk attitudes, Agarwal and Ramasnani (1992) found that risk-taking agents are usually better able to enter export markets. In our research, the farmers' risk attitude is significant, but the direction of the sign is opposite: risk-averse agents are more likely to deliver to the export market. This seems to be a counter-intuitive result, but producers who are delivering to the export market at this moment in Costa Rica are to a certain extent playing safe by reducing transaction costs and delivering to the market outlet that is closest to their orchard, even if they have to comply with more requirements to have access to such an outlet, not only in terms of product quality but also in terms of increase production costs.

Experience in crop production also appears to be essential to be able to enter the export market, but in the case of mango experience it has a negative sign, thus implying that less-experienced agents are better able to engage with the export market. Saenz-Segura (2006) and Saenz-Segura and Ruben (2004) found a contrasting relationship. The fact that mango exports from Costa Rica are a rather new activity and that there has been a shift to new varieties of mangoes makes it is easier for new (but less experienced) growers to adopt the production system required for export. Established producers are more likely to deliver to the local market, and are equally less risk-averse given the trading options in the local market.

2.4.4 Production system

The variables of cultivated mango area and scale of production have a positive relationship with export market orientation, in line with the findings from the literature (e.g. Agarwal and Ramasnani, 1992; Agarwal and Gort, 2002; Saenz-Segura and Ruben, 2004). Larger mango areas increase the possibility for producers to deliver to the export market segment; likewise, a higher scale of production (more boxes per week) tends to facilitate the entrance into the export market. From the institutional point of view, it is important for the buyer to contract a limited number of larger producers, thus reducing the monitoring and control cost as well as the delivery uncertainties. Key and Runsten (1999) pointed out that business relations between large companies or vertical integration of processes are common procedures for reducing transaction costs associated with uncertainty (quality, control and monitoring).

2.4.5 Price attributes

Price differences seem to have limited implications for the market outlet choice, since mango producers are mostly price-takers and are not able to influence prices that are out of their control. Mango prices in Costa Rica are based on the international baseline price provided by the importers to the local producers and the bottom price they receive in the local market. This price transmission mechanism is especially valid for the period between January and May when the mango export window is open for Costa Rica. Mango prices are also dependent on other fruit categories, since (local) consumers are willing to use different varieties of mango and other tropical fruits depending on seasonal availability.

2.4.6 Market context

From the institutional point of view, purchase conditions related to market entry barriers as well as to contract requirements are of critical importance for entering the export market (Key and Rusten, 1999; Papadopoulos *et al.*, 2002; Chiarelli *et al.*, 2000). Farmers that are better able to comply with the demanding market conditions will become engaged with export outlets. In the case of the Costa Rican mango sector, export producers are agents that are willing and able to tolerate stricter and more demanding delivery conditions for entering the market. In addition, the distance to the market where mango is collected facilitates better quality control and reduced degradation. Farmers must construct recollection facilities close to their production areas in order to improve the production systems oriented towards the export market,

In Costa Rica, mango production for the export market is based in Puntarenas (in the Western part of the country), while production for the local market is mainly located in Alajuela (in the Central part of the country). Hence, export market collectors and packaging facilities are closer to the mango producers in Puntarenas, whereas producers delivering to the local market must travel longer distances for access to the local markets found mainly in the Central Valley where most of the population is concentrated.

Purchase conditions for export-oriented mango are different from those for the local market oriented fruit. The quality standards are stricter as are contract characteristics such as contract type, method of payment, rejection rate tolerance and supervision by the buyer in the orchard. Normally, exporter-oriented growers can tolerate more rejection in the selection and packing process, must sign written contracts and have less flexibility in their selling strategy. Local market oriented producers have less requirements to satisfy customer needs and have a more flexible selling strategy.

2.4.7 Differentiation within mango supply chains

Finally, we can discuss the behavioural attributes that give rise to differentiation within the mango supply chain. Therefore, we use an ANOVA analysis to identify the contractual delivery conditions prevailing in transactions between the different actors in the chain. The diagram in Table 2.4 summarises the main factors affecting the delivery conditions at the interfaces between major market agents.

Table 2.4. Delivery procedures at supply chain interfaces (ANOVA analysis).

	Retailer	**Intermediary**	**Exporter**
Wholesaler	(-) Number of buyers	(-) Method of payment	(+) Years attending the market
	(-) Price knowledge	(+) Distance to market	(-) Type of agreement
		(-) Buyer visits the orchard	(-) Method of payment
			(-) Buyer visits the orchard
			(-) Supervision by the buyer
			(-) Quality control
			(-) Access to credit
Retailer		(+) Number of buyers	(+) Number of buyers
		(+) Years attending the market	(+) Years attending the market
		(-) Rejection rate	(-) Rejection rate
		(-) Method of payment	(-) Type of agreement
		(+) Distance to market	(-) Method of payment
		(+) Price knowledge	(+) Distance to market
		(-) Buyer visits the orchard	(-) Buyer visits the orchard
			(-) Supervision by the buyer
			(-) Quality control
			(-) Access to credit
Intermediary			(-) Rejection rate
			(-) Type of agreement
			(-) Buyer visits the orchard
			(-) Supervision by the buyer
			(-) Quality control
			(-) Access to credit

Major differences between delivery conditions at the wholesale market (Cenada) and the retail market (farmers' market) refer to the larger number of buyers and the better availability of price information in the latter market. This implies that local retail outlets provide more bargaining opportunities for producers. For deliveries to the wholesale market, producers must travel longer distances, but they receive direct cash payments from buyers and are not bothered by buyers wanting to inspect their orchards. At this interface, rejection rates might be higher, but options for selling a larger volume and the absence of delayed payment risks may make wholesale delivery an attractive option, particularly for larger producers. The main difference between export and wholesale outlets is that an export outlet implies a more complex delivery agreement, payments by cheque, and more field supervision and quality control, but also opportunities for additional access to input credit as part of an interlinked contract.

Delivery conditions at retailers are differentiated from those with intermediaries with respect to the larger number of buyers, greater distance to the market and the longer period attending the market. This results in a lower rejection rate, direct cash payment and more knowledge about prices but less information on the production conditions for deliveries to the open-air farmers' market. Retail sales can be expected to be attractive for producers that offer standard mango quality and prefer prompt sales with too much risk. Otherwise, sales to intermediaries are likely to be preferred by producers with above-average mango quality that can sell the rejected fruit through alternative market outlets.

Delivery modes at the export market are mainly based on the contractual arrangements where producers are willing to bear stricter conditions in order to get access to export outlets. Given their remote location, most of these producers only have limited alternative options. In case they do not agree to sell to the export agency, they have to face high transport costs to reach other outlets or deliver elsewhere to intermediaries that will pay substantially lower prices for the mangoes compared to the export market. The main differences between the intermediary and the exporter include more complex delivery requirements and higher quality controls that are in effect in the export market segment.

We can now outline the mango supply chain configuration and the differences in outlet choice (see Figure 2.2.). It is important to observe that for the export market vertical integration processes are already quite advanced. Producers deliver on the demands of the buyer and therefore face higher rejection rates, but in compensation they have access to stable market outlets, receive input and (subsidised) credit and can benefit from lower transport and delivery costs. For the local market, the producers' experience and their historical knowledge and relationships with the market are of key importance. In addition, the flexibility to sell to a large number of different clients, the cash payments and the reduced possibilities of hold-up may make this a preferred market outlet. Sales to intermediaries that live close to the producer and visit the producer often to see whether rejected mango can be bought provide an important secondary market outlet. Since local markets are not as strict with respect to quality (but do control colour and maturity), intermediaries purchase mangoes that are riper than those

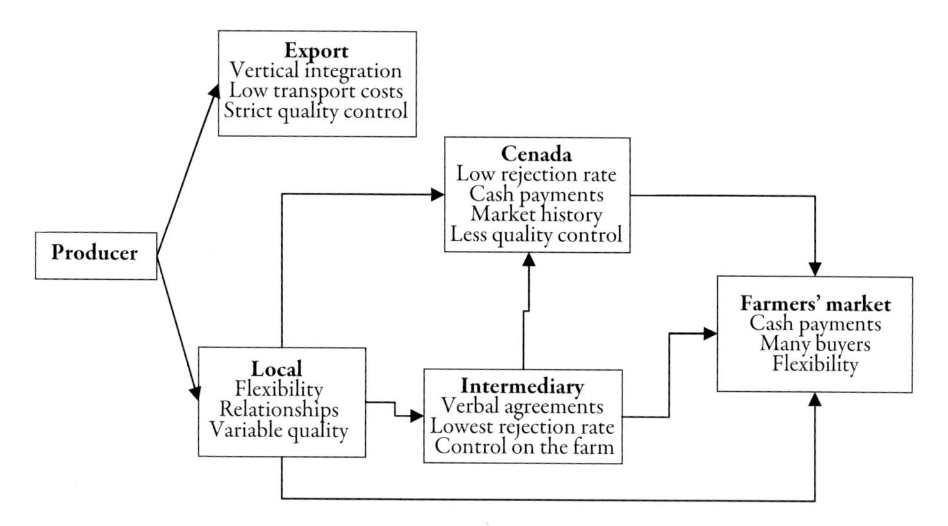

Figure 2.2. Differentiation in mango outlet choices.

delivered to the export market. However, intermediaries buying mango usually pay one week later (after having sold the produce), and therefore they build on an established relationship with the producer, whereas for Cenada and farmers' market sales, spot market transactions are the rule.

2.5 Discussion and conclusion

Mango market outlet selection is a strategic decision for producers, who have to decide on their market portfolio mix in terms of maximising their (family) welfare and guaranteeing their strategic market position. Since price differences between local and export markets are not substantial, other market delivery conditions related to guaranteed and stable access and potential cost advantages tend to be more important. While deliveries to the export market might be attractive if producers can benefit from reduced transport and transaction costs, they also incur higher input costs, have to face higher rejection rates and must pay fees for their Eurep-GAP certification. This implies that all production activities should be registered and that producers lose their independence, since they have to accept *in-situ* inspections. Therefore, many producers moved to the local market because they were unable to meet the certification requirements.

Given the variability in prices (at local and export markets) and the additional differences in production conditions (input and credit costs) and supply conditions (rejection rate), market outlet choice is a complex decision involving both welfare objectives and risk and transaction costs considerations. Although producers delivering to the local farmers' market may sometimes receive higher prices, they also face higher costs for delivering their mangoes to this outlet, and they need specific skills and knowledge to guarantee successful operations. Building reputation and trust requires experience and a long-term engagement. Otherwise, direct cash payments at the retail and wholesale market tend to reduce risks. The operations of local intermediaries are somewhat closer to the producer and enable deliveries of rejected produce to an alternative market outlet. The inspections of the orchards by intermediaries are thus not related to quality control so much as availability control. Compared to exporters, intermediaries face major financial constraints for direct payments, and usually take advantage of opportunism in their pricing policy.

Producers delivering to the export market have to tolerate more stringent delivery conditions as defined by the buyer. Export market chains are moving in the direction of vertically integrated processes. Given the risk involved in mango production, exporters will contract with producers instead of engaging in production themselves. Export contracts are mainly arranged with medium-scale and larger mango orchards that use up-to-date procedures and cultivate modern varieties, and are located near the processing plants. In the export chain, producers are the least informed about market conditions and opportunities, and are mainly concerned with upgrading primary production. Traders and intermediaries are the best-informed agents in the chain since through their logistic operations, they link the production and marketing part of the chain.

This research presents a standard analysis of market outlet choice, but contributes to the idea of comparison among outlets as well as the view of complementarities of markets at least for the mango chain in Costa Rica. All the results of the analysis are in line with previous research, except for household characteristics, such as experience and risk attitude, which are contrary to what was expected. The findings show that in order to promote mango producers to choose the export market, they must be young and risk adverse, most likely because it is easier to adopt new production systems and rules of the game than change the old way of doing business. Further research on other non-traditional commodities must be conducted to understand the most significant household characteristics for promoting exports among other things, and must thus be able to define incentives for different types of producers.

Further research will be required to obtain detailed data on the production function for different market outlets. It would also be important to determine the market mix of the producers to be able to understand their multiple market orientation. Both aspects could provide further information on the backwards implications of market outlet choice for the strategic behaviour of mango producers.

Chapter 3. Economic incentives for improving quality in the mango sector of Costa Rica[5]

3.1 Introduction

Mango (*Mangifera indica*) has been cultivated as a food crop for more than 40 centuries. The production originated in India and Malaysia and has spread to tropical and subtropical countries around the world. There are currently some 150 varieties of mango known worldwide, but only a few (i.e. Haden, Irwin, Keith, Tommy Atkins) are of commercial importance. Mangoes were first imported to Costa Rica in 1796 from different Caribbean countries (Lezema, 1989). In the early 1970s, the University of Costa Rica introduced some special red and yellow varieties in the Orotina area for commercial purposes (Mora, *pers. comm.*, 2004; Mora *et al.*, 2002). At the beginning of the 1980s, the production of red mango varieties for (European) export markets became more important, leading to a further reduction of the local cultivated area of yellow mango (Jirón, 1995; Buzano, 1997).

Mango quality varies greatly due to differences in taste, flavour, colour, aroma and size. Consumers look for mangoes with no external damage, with a stable weight, colour and consistency, and at a reasonable price. Producers' crop management and post-harvest practices as well as delivery systems have a profound impact on observable and imminent quality characteristics (Ruben *et al.*, 2005). Although a fair amount of knowledge is available regarding the technological options for improving mango quality (Garvin, 1984; Montero and Cerdas, 2000), there is far less understanding of which economic incentives and contractual regimes might be effective for providing producers with the necessary incentives for adjusting their mango production and management systems in order to enhance quality performance in line with consumer demands.

In this research we use an innovative analytical framework to link consumers' quality perceptions with producers' quality management practices, and the impact of different producers' characteristics and marketing arrangements with improved quality performance for specific local and international market outlets. We collected detailed data on farm households and mango production systems from a sub-sample of 35 mango farmers in the five main production zones of Costa Rica. In addition, quality parameters of five selected mango fruits were assessed according to eight different criteria (mangoes were checked on weight, various types of damage, irregularity of the shape, colour of the peel, firmness of the pulp, colour of the pulp, total soluble solids – TSS and pH). They were subsequently weighted by a consumer panel and by using trader expert opinions. A mango quality index is specified

[5] An earlier version of this paper was submitted for publication to the International Journal of Quality and Management Reliability (IJQMR).

for both local and export market outlets. Finally, we identify the relationships between farm-household characteristics, production system features, marketing relationships and intrinsic mango quality attributes.

The remainder of this chapter is structured as follows. First, we provide an overview of the literature dealing with quality and incentive linkages and we provide a general analytical framework to examine the factors that influence quality performance. Thereafter, we analyse the structure of the mango chain in Costa Rica and present basic data on producers and product samples. Next, we outline the procedure used to construct the quality index for mangoes and apply this index to differentiate between the local and international market segments. The quality index is subsequently related to particular farm-household characteristics and a number of price and non-price factors that are part of the delivery contracts with traders. We conclude with some implications for policies and indicate options for further research.

3.2 The mango supply chain in Costa Rica

The main mango varieties grown in Costa Rica suitable for export are Tommy Atkins, Kent, Keith, Palmer and Smith (Mora, *pers. comm.*, 2004; Jiménez, *pers. comm.*, 2003; Central Pacific Census, MAG, 2004). In 2003, mango orchards in Costa Rica covered 8,350 ha and the quantity produced reached 35,000 tons. In seven years (1996-2003), the area went up by 7% and the amount produced rose by 75% (Jiménez, 2003). This increase was due to technological change (Montero and Cerdas, 2000). Costa Rica has about 1,950 mango producers, 60% of which cultivate less than 5 ha, 35% cultivate between 5 and 20 ha, and 5% own more than 20 ha (Mora, *pers. comm.*, 2004; SEPSA, 2001). The producers are organised in different ways; large and medium-sized producers are linked to international trading companies, and small producers are either organised in co-operatives or producers' associations or not organised at all.

Figure 3.1. shows the main locations of mango producers in Costa Rica. The circles on the map represent the mango production area and the area where the fieldwork for this research took place.

The organisation of the mango supply chain is relatively simple for the export market and much more complex for the local market (see Figure 3.2.). For the export market there is a straightforward relationship between the traders and the producers, but at the local market there are many different intermediaries involved. These agents play many roles in the chain: they buy mangoes from the producers, sell the produce to the Cenada (wholesale market), or buy from Cenada to deliver to local outlets such as retailers and the wet market. Hortifruti is a private company that buys directly from the producers and sells to the main supermarkets in Costa Rica.

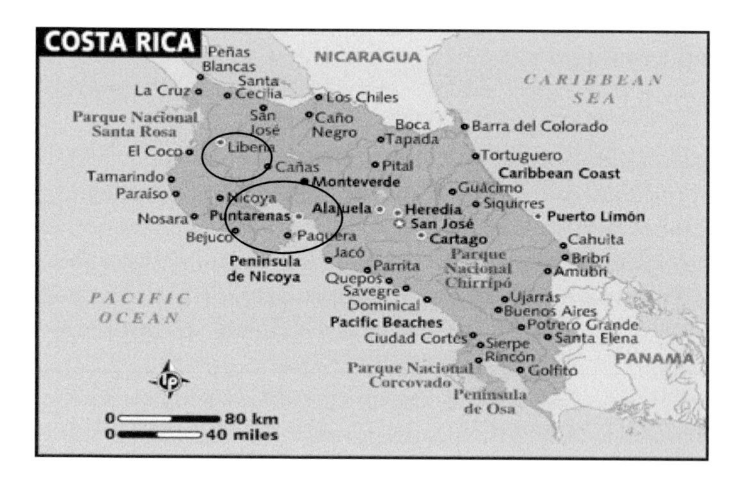

Figure 3.1. Main mango producing areas in Costa Rica.

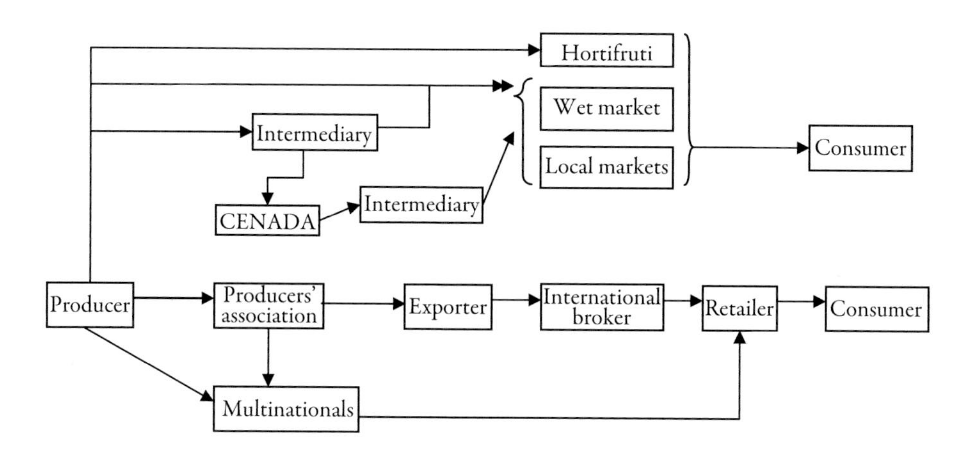

Figure 3.2. Mango supply chain in Costa Rica.

3.3 Quality perceptions and incentive regimes

Food quality is a broad concept that has been defined in a variety of ways, using either objective or subjective criteria. While the theoretical relationship between quality perceptions and consumer choices is widely addressed in the literature (see: van Trijp and Steenkamp, 1998; Sloof *et al.*, 1996), there is surprisingly little attention given to the incentive regimes that effectively enable producers and processors to respond to these market demands. We therefore aim to develop an integrated procedure for linking intrinsic quality attributes as perceived and valued by consumers with producers' characteristics that influence quality performance, in order to identify which price and non-price elements are able to dovetail farming systems management systems with market criteria.

Different definitions of quality are put forth in the literature. Juran (1990) defines quality as product performance that results in customer satisfaction and freedom from deficiencies in order to prevent customer dissatisfaction. Crosby (1979) describes quality as compliance with clear product specifications, whereby the firm management is responsible for maintaining univocal attributes. A more comprehensive definition is provided by Luning *et al.*, (2002), who refer to quality as the ability to meet or exceed customer expectations.[6]

The different attributes included in the concept of quality depend on the relevant actor who is acquiring the product. Major actors participating in the valuation of food quality for the export market include producers, processors, exporters, importers, wholesalers, retailers and consumers (see Figure 3.3.); external agents such as voluntary agencies and the government may also influence these perceptions. Wholesalers and retailers emphasise visual attributes such as size, shape, colour and shelf life, taking into consideration consumer preferences. However, consumers are interested in many more aspects related to food quality, such as taste, freshness, appearance, nutritional value and food safety. Government officials are involved in regulations concerning health and safety aspects. Producers and processors commonly prefer

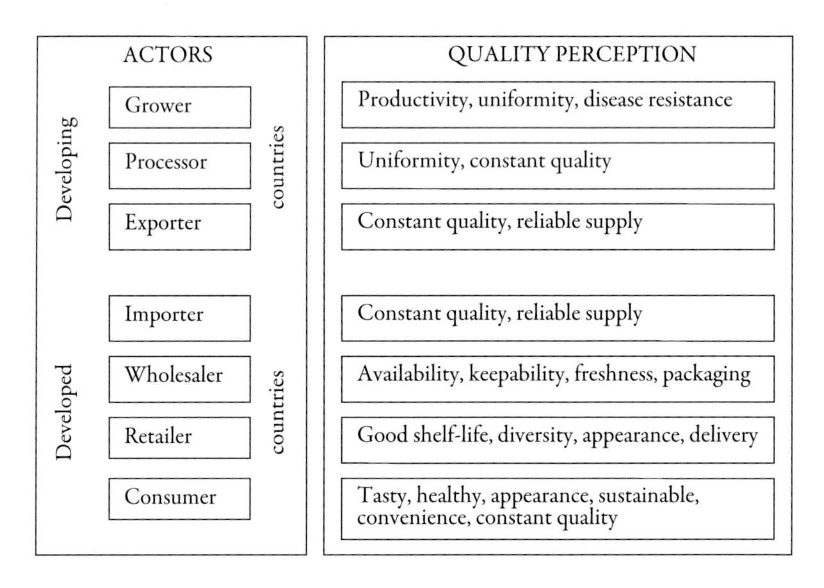

Figure 3.3. Actors in the mango chain for the export market and their quality perception.

[6] The customer can be any actor in the chain, with the consumer as the final customer.

profit attributes, including higher yields, suitability for mechanical harvesting and industrial preparation, and resistance to pests and diseases.

The quality of agri-food products can be understood as an aggregate of various quality characteristics, which can be divided into intrinsic and extrinsic quality attributes (van Trijp and Steenkamp, 1998; Steenkamp, 1989; Poulsen *et al.*, 1996). Intrinsic quality attributes refer to physical product characteristics and can be measured in an objective manner and by sensory perception. The combination of all these attributes determines the intrinsic product quality. Extrinsic attributes refer to how the food is produced. These extrinsic factors commonly have no direct influence on the observable characteristics of the product, but they can be of overriding importance in the purchasing decisions of some consumers.

Fruit producers tend to focus on the quality attributes that can be observed directly. It is rather common for mango producers in Costa Rica to demonstrate quality by comparing the mangoes from their own orchards with those of neighbouring farms. Normally, another way to benchmark quality is by providing information about the technological package and the crop management procedures used on the farm. Garvin (1984), however, emphasises that farmers should be more objective when defining the quality performance attributes of their fruit and relating them to their production practices.

Sloof *et al.* (1996) provide an extensive literature review and have identified four different types of approach to quality:
- The *transcendent approach*, in which quality is considered a property that an individual can learn to recognise only through experience;
- The *production-based approach*, where it is dealt with during the production, distribution and commercialisation process. The required quality is met when a product complies with the standards (specifications) for the next step in the chain;
- The *value-based approach,* where quality is a composite of product characteristics that are valued in the market and enable producers to differentiate between their products to maximise profits and reduce competition;
- The *user-based approach,* which places the user in the centre; the quality of a product is based on how the end-user perceives it, and as such depends on the individual user's subjective perceptions, needs and goals.

These broad definitions, dimensions and attributes of quality take us to the question of how to link user-based characteristics with production-based approaches in such a way that quality is understood as a physical product property and related to the production-system characteristics and the management of the supply chain as well. We therefore relied on a techno-managerial approach (Luning *et al.*, 2002) that identifies the product and process variables influencing quality management practices and consumers' preferences.

Randall and Sanjur (1981) and Gains (1994) explain that food preferences are related to three sets of key variables: 1) the characteristics of the individual, 2) the characteristics of the product and process, and 3) the characteristics of the market and institutional environment. Consequently, Sijtsema (2003) concludes that food perception models for product development should take into account the environment, the individual, the food and the context of consumption. In a similar way, Kahn (1981) further has disaggregated food preferences into seven factors: 1) personal factors, 2) socio-economic factors, 3) education, 4) cultural, religious and regional factors, 5) intrinsic factors, 6) extrinsic factors, and 7) biological, physiological and psychological factors. Given these multiple aspects, quality tends to be a composite index of various attributes, and it becomes extremely difficult to identify which incentives can be used to influence quality performance.

3.4 Analytical framework

We explore the most important intrinsic quality attributes of mangoes in relation to the socio-economic features of the mango producers. First, a detailed characterisation of the producers and of different types of contractual arrangements for mango was carried out. Contractual aspects refer to non-price criteria used by traders/processors to select producers that deliver the desired quality. Key contractual attributes that are considered refer to trust relationship with the buyer, bargaining power, quality compliance and the willingness to sell the produce. Food quality management is a complex system, since we have to consider both human and food behaviour (Luning *et al.*, 2006), where food behaviour is a function of food dynamics and technological conditions, and human behaviour is a function of individual dynamics and administrative conditions. In the present research we intend to analyse quality management of mangoes using a similar approach, but adding price and non-price attributes to the analysis.

In the following analytical framework (see Figure 3.4.), we explain differences in the (intrinsic) quality of mango in relation to the socio-economic characteristics of the household and the technical management of mango orchards. Hereafter, we identify the price and non-price attributes that are effective in influencing quality performance.

To make this basic model operational, we need to understand the relationships between quality performance, farm-household characteristics, production system organisation, delivery conditions and contract design (i.e. non-price attributes) and price incentives. Quality of crop production will be affected by different farm management practices (Ali, 1995; Odulaja and Kiros, 1996). Such managerial practices include both physical inputs (land size, implements, etc.) and the farmer's personal endowments (age, literacy level, and experience). One of the important management practices is lean production.[7] Most common benefits related to lean practices are the improvement of labour productivity and higher quality performance (White *et al.*, 1999). Empirical evidence regarding the implementation and adoption of lean practices

[7] Lean production is a multi-dimensional approach encompassing a wide range of management practices (including quality systems) to streamline the integration of the supply chain (Shan and Ward, 2000).

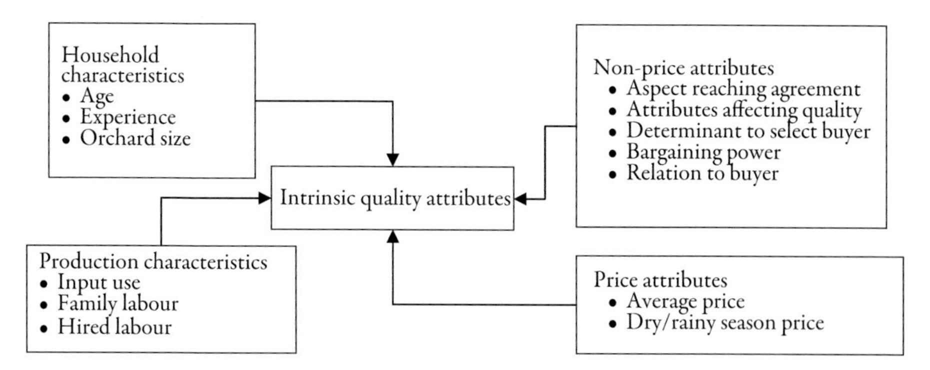

Figure 3.4. Analytical framework. Intrinsic quality relationships to socio-economic variables.

points towards a positive impact of firm size: large firms are more likely to implement lean practices than smaller ones (Shan and Ward, 2000),[8] and therefore produce higher quality products. Earlier experience in the activity is usually related negatively to the implementation of lean production, since it would be harder for experienced farmers to adjust management practices (Pil and MacDuffie, 1996). Just and Zilberman (1983) find that technology change and quality improvements are mainly adopted by large farms.[9]

Suitable incentives for enhancing the adoption of quality innovations must be identified to enable supply chain agents to adjust their production practices and management regimes accordingly. Both prices and non-price attributes might be helpful for this. Zeithaml (1988) suggested that in the market the 'get' and 'give' component of goods or services should to be matched. Given this definition of value, a trade-off between quality and price might arise (i.e. value-for-money). It is generally acknowledged that higher quality produce is rewarded with a higher price (Batt and Parining, 2000). Menegay *et al.* (1993) have shown, however, that variations in quality can sometimes lead to lower prices, since farmers sell their production without any grading to collecting agents and distributors who purchase the entire crop, irrespective of the quality.

[8] Studies of goat production in South Africa reveal that farms with a smaller herd face high transaction costs to implement quality improvements, since the input required for enhancing quality (i.e. medicines, ear tags, laboratory tests and transport equipment) can be prohibitively expensive (Roets and Kirsten, 2005).

[9] It is very common in mango production in Costa Rica to lease land from subsistence farmers to increase farm size and production. These farmers, however, care less about land conservation and use resources in a more extractive way, thus affecting the sustainability of the resource base.

Contracts are an alternative device for marketing the fruit. The literature suggests that vertical integration[10] could lead to increased quality due to standardisation of production, harvesting and handling practices. Global supply has brought with it new demands by retailers for traceability, tractability, quality assurance and consistency of deliveries. The consumer expects to find the product of a specified quality at a certain price and with a guaranteed shelf life. These demands have led to an increased occurrence of vertical integration in agricultural food and non-food markets (Sporleder, 1992; Martinez *et al.*, 1997; Gow *et al.*, 2000; Engelbart *et al.*, 2001; SAMIC, 2002; Singh, 2002). Sáenz-Segura (2006) found that the existence of a delivery contract has a positive effect on the quality of the product. Alexander *et al.* (2000) stressed that the quality of tomatoes will be higher in the presence of (incentive) contracts. For these reasons, we have considered incentives through contracts to be more appealing to the mango producers than other incentives are. Mango producers from Costa Rica are price-takers since mango prices are defined by the international market. Hence we argue that producers' behaviour can be modified by the implementation (adding or eliminating) of certain contract clauses and improving relationships and coordination, and not only by produce price incentives.

3.5 Materials and methods

Our framework for this analysis is based on three components. First, we construct a quality index for mangoes based on product attributes valued by consumers (using a consumer panel). This index is further disaggregated for relevant local and export market outlets. Second, we identify the socio-economic and technical characteristics of the mango producers and their production systems that influence management regimes. Third, we construct some relevant behavioural factors that characterise market outlet choice and the contractual relationship between producers and buyers. The final analysis concerns the interaction between producers and contractual dimensions with quality performance.

3.5.1 Data

We collected three types of data from a random sample of mango producers. Data collection includes the measurement of mangoes for quality inspection, a questionnaire completed by producers regarding farm-household and production characteristics, and a series of interviews with producers to reveal their preferences regarding outlet choice and relationships with traders. The entire sample includes 94 farmers selected randomly from the producers' census for the Central Pacific zone (CPZ) of Costa Rica, conducted by the National Production Bureau (CNP) in 2003.[11]

[10] Singh (2002) explained that foreign direct investments foster vertical integration between suppliers and the food industry and could enhance farmers' access to basic inputs, credit, and market information, to ensure timely delivery and high-quality supplies. Vertical integration and contract farming are thus likely to increase the quality of the product delivered to the market.

[11] See Box 2.1.

Primary data regarding the quality of mangoes for the domestic and export market was collected amongst a sub-sample of 35 mango producers sampled at five producing regions, all located in the Central Pacific zone of Costa Rica during April and May 2005. The quality measurements carried out are based on the most relevant intrinsic quality attributes: appearance (weight, irregularities and external damage), texture (firmness), colour (peel and flesh) and indicators for taste (Brix, pH). The quality attributes selected are indicators for ripeness of the mango. From each producer, five mangoes were randomly sampled mostly after harvest and sometimes directly from the tree. The samples consist of seven different varieties of mangoes.[12]

3.5.2 Quality index

The major problems presented by the mango are a deformed shape, damage caused by the tree, and damage by bugs, fruit flies, and latex flecks. These problems are mainly related to poor post-harvest practices, such as tree cutting, pest and disease control, fertilisation and irrigation (Montero and Cerdas 2000). Mango quality is also strongly influenced by harvesting methods and timing and is determined by: 1) the shape of the fruit (the 'shoulders' should be full and higher than where the stem attaches to the fruit), 2) the colour of the skin, 3) the colour of the pulp, 4) the firmness of the fruit, 5) the specific weight (considering that mangoes with a volume of more than 10% water are not suitable for export), and 6) the number of days since the tree began to bloom (between 93 and 115 days, depending on the cultivar).

Molnár (1995) describes the concept of food quality as: 'the quality of food products in conformity with consumer requirements and acceptance, (....) determined by their sensory attributes, chemical composition, physical properties, microbiological and toxicological contaminants, shelf life, packaging and labelling.' He designed a model for the overall assessment of food quality based on multiple attributes, taking into account that separate attributes of quality interact heavily and that this inter-correlation cannot be ignored.

For the operationalisation of a measurable indicator of food quality we need to consider not only the properties of the food product, but also the consumers' perception of these properties. In order to make quality more tangible, a division should be made between the intrinsic and extrinsic quality attributes. A hypothetical quality function Q has been proposed for food products (van Boekel, 2005) (Equation 3.1):

---

[12] The mangoes collected were stored for one night at room temperature and were analyzed the next day at the food chemistry laboratory of the University of Costa Rica (San José, Costa Rica). The fruits were weighed and judged for the various external attributes (irregularity, colour of peel, damage). The texture of the fruit was measured with a texture analyser (TA XT Plus, Stable Microsystems). The colour of the pulp (L*, a*, b*) was analyzed with a colorimeter (Colorflex CX1192, Hunterlab). Juice was obtained with a juice extractor for the analysis of pH (Mettler pH-meter) and total soluble sugars (TTS, indicating sweetness) using a digital refractometer (K71601, Optech Technologies). A subjective descriptive scale was used for the assessment of overall appearance (from 1 to 5; colour of peel and irregularities) and damage (from 0 to 5; fruit fly, brown spots, stem-end rot, internal damage, scars and bruises).

$$Q = f(Q_{int}, Q_{ext}) \tag{3.1}$$

where quality Q is a function of intrinsic and extrinsic quality attributes that can be decomposed into intrinsic and extrinsic quality functions Q_{int} and Q_{ext}. The functional form of this function is still undefined. A further decomposition of the intrinsic quality function Q_{int} into a series of intrinsic quality attributes I_i is possible (Equation 3.2):

$$Q_{int} = f(I_1, I_2, \ldots, I_n) \tag{3.2}$$

Molnár (1995) divides the attributes composing total quality into five groups: 1) sensory attributes; 2) chemical composition and physical properties; 3) microbiological and toxicological contaminants; 4) packaging and labelling; and 5) shelf life. Our analysis is mainly focussed on sensory attributes and physical properties. We used the following seven quality attributes in the analysis: weight, sweetness (TSS brix), acidity (pH), firmness, colour of skin and pulp, external damages, and irregularities in shape.

In this research we have focussed on the directly measurable intrinsic aspects of food quality. Since intrinsic quality attributes may be complex to handle, we have used the quality attributes as performance indicators. For instance, nutritional value can be considered the result of acidity that is indicative for vitamin C content. Within each category, relevant attributes are identified based on the knowledge of experts (Delphi method). Each attribute I_i is ranged into the parameter Z_i with $1 \geq Z_i \geq 0$. Z_i has a value of 1 when I_i is at its optimum value, and the value 0 when I_i is at its lowest and becomes unacceptable. Therefore, I_i represents the evaluation of the attribute and Z_i the appreciation of the attribute. To obtain a total quality, each attribute is assigned a weight and the attribute categories are also assigned weights. This results in the following mathematical equation for the overall quality index Q (Equation 3.3):

$$Qi = W_1 \sum_{j=1}^{k} w_{1j} \cdot z_{1j} + W_2 \sum_{j=k+1}^{l} w_{2j} \cdot z_{2j} + W_3 \sum_{j=l+1}^{m} w_{3j} \cdot z_{3j} + \ldots \tag{3.3}$$

where

$Qi =$ overall quality index (minimum of 0 maximum of 1);

$W_1, W_2, W_3 =$ weighting factors for groups of attributes whose sum is equal to 1 for each food product;

$w_{1j}, w_{2j}, w_{3j} =$ weighting factors for attributes within the same group, whose sum is equal to 1 for any food product within the same group;

$z_{1j}, z_{2j}, z_{3j} =$ values ranging between 0 and 1, according to acceptance;

$k, l, m, n, \ldots =$ number of attributes.

Molnár (1995) suggests adding up weights for the quality function as judged by experts. This could help to disentangle intrinsic quality attributes. However, consumers[13] are the ones who finally assign a quality to the product according to their own priorities. For this reason, here in addition to the expert's judgment we have used a consumer panel[14] that evaluated and weighted the various quality attributes relevant for mangoes. In this way, the objective as well as the subjective quality was used to assess the overall quality. The most relevant clusters of mango quality attributes used in this study are:

- 'Damage', composed of the attributes related to internal and external damage cause both by nature and by humans;
- 'Maturity', composed of the attributes related to *ripeness* including sweetness, acidity, firmness, colour of skin and pulp;
- 'Appearance', including the attributes of weight and irregularities in the shape;
- 'Variability', related to the variability between mangoes from the different producers.

This analytical procedure further enables us to disaggregate the quality index that we have derived according to specific quality attributes. Hence we can specify one quality index, which is defined as follows (Equation 3.4):

$$Qi = W_1 \sum_{j=1}^{k} w_{1j} \cdot z_{1j} + W_2 \sum_{j=k+1}^{1} w_{2j} \cdot z_{2j} + W_3 \sum_{j=l+1}^{m} w_{3j} \cdot z_{3j} + W_4 \sum_{j=m+1}^{n} w_{4j} \cdot z_{4j} \quad (3.4.)$$

where

Qi =	Quality index;
W_1, W_2, W_3, W_4 =	weighting factors for damage, appearance, maturity, and variability of attributes whose sum is equal to 1 for each food product;
$w_{1j}, w_{2j}, w_{3j}, w_{4j}$ =	weighting factors for attributes within the same group, whose sum is equal to 1 for any food product within the same group;
$z_{1j}, z_{2j}, z_{3j}, z_{4j}$ =	values ranging between 0 and 1, according to acceptance;
$k, l, m, n, \ldots$ =	number of attributes.

After presenting this function, the study focuses on the relationships between the intrinsic quality attributes and the socio-economic characteristics of producers as well as the determinants of contract choice.

[13] We know that consumers in general are not homogeneous, but for this study we did not differentiate between consumers. We expect to have different consumer preferences from local and international consumers, and segments with different preferences could be identified within those broad groups of consumers.

[14] A sensory panel has not been used by Molnár, but it is believed that the input of consumers makes the model more realistic, and more applicable to the local market. The participants of the consumer panel were asked to rank some mango characteristics from most important to least important. They were also given pieces of mango and asked to indicate how much they liked the piece of mango on three scales. The same mangoes were later analysed in the lab to correlate the preferences with technical data.

3.5.3 Farm-household characteristics

From the socio-economic survey amongst mango producers, we measured a set of relevant farm-household characteristics (see Appendix A). These characteristics are related to both quality indices (Equation 3.5):

$$Qi = f\ (\chi_{ij}) \tag{3.5}$$

where χ_{ij} represents household characteristics and production system features. The following variables are considered: age and experience with mango production (in years), use of hired labour (cost per year), family labour use (cost per year), input use (cost per year), mango area (ha), rainy season price (average coloness per season), dry season price (average colons per season), average price (average colons per year), production (kg), and the perception of bargaining power (1 to 10 on the Likert scale).

Categorical regression (CATREG) is applied for quantifying categorical data by assigning numerical values to the categories, resulting in an optimal linear regression equation for the transformed variables. CATREG extends the standard approach by simultaneously scaling nominal, ordinal, and numerical variables. The procedure treats quantified categorical variables in the same way as numerical variables. Using non-linear transformations allows variables to be analysed in various combinations to find the best-fitting model.[15]

3.5.4 Contract choice

The final stage of this analysis focuses on the identification of relevant behavioural variables that influence quality performance. Therefore, we performed a factor analysis (FA) based on producers' perceptions regarding their relationship with the buyers. We have calculated the following functions (Equation 3.6):

$$Qi = f\ (y_{ij}) \tag{3.6}$$

where y_{ij} represents composite factors reflecting behavioural characteristics. The following factors are derived: critical attributes for reaching an agreement, attributes affecting product handling, determinants for selecting a specific buyer, bargaining power and affective relationship with buyer. All of these constructs are perceptions measured on a Likert scale ranging from one (=low) to ten (=high). Each factor is analyzed separately against the quality index and finally all factors are run together to obtain a joint solution.

[15] The main assumptions for CATREG are that only one response variable is allowed, but the maximum number of predictor variables is 200. The data must contain at least three valid cases, and the number of valid cases must exceed the number of predictor variables plus one. We have 35 cases and the analysis was conducted for one variable at a time against the variable of the quality index. Subsequently, the analysis was carried out for two or three variables at a time against the quality index. This analytical process enabled us to respect the basic assumptions of the technique.

3.6 Results

Below we present the results regarding the characterisation of producers and their production systems and the measurement of the quality index, followed by an assessment of the impact of structural and behavioural factors on quality performance.

3.6.1 Socio-economic characteristics of mango producers

The socio-economic characteristics of 94 mango producers from five different locations in Costa Rica are presented in Appendix A. The mango population is very heterogeneous, not only in terms of household characteristics but also in production system characteristics. Family size varies from 3.2 persons in San Mateo up to 5.3 in Orotina; in San Mateo the average age of the head of the family is 36.5 years, whereas in Esparza it is 50.8 years. Orotina producers have the most mango experience, whereas those in Esparza have the least experience. Producers from San Mateo and Orotina have the largest mango orchards (> 20 ha). Orotina and San Mateo have the highest specialisation rate with more than 75% of the farm area devoted to mangoes, followed by Puntarenas with 65%, and the rest of the locations use less than 50% for mangoes.

Producers in Esparza rely mostly on hired labour, while in San Mateo far more family labour is used in production and harvesting. In general terms, producers who hire most of the labour find it difficult to enforce quality control. Orotina is the only area with losses, since producers receive a lower average price. This may be caused by the high concentration of mango producers. In Garabito the highest yield (6.65 t/ha) is obtained, while the yields are lowest in San Mateo (5.08 t/ha). All locations selected have yields well above the national average of 4.2 ton per hectare.

3.6.2 Quality index

In Table 3.1 we present the different values obtained in the measurement of quality attributes. The first two columns include the quality index for local (*QiLM)* and export markets (*QiEM)*. Differences between the two indexes are mainly due to the different quality requirements in both markets. It is evident that the export market is considerably more demanding.

In Figure 3.5., we present the results of the QiEM analysis of individual producers, where one of the straight lines represents the minimum intrinsic quality for the export market (defined by the use of the minimum export quality requirements and defined by the consumer panel and the interviews to importers and exporters) (see Box 3.1). There is a complete separation among the mangoes showing the quality differentiation between markets, although some mangoes meant for the local market have the intrinsic quality attributes to be sold in the export market.

Table 3.1. Average results of chemical and physical quality analysis by region.

Location	**Quality index**		**Weight (g)**	**Irregularity (scale 1-5)**	**Damage (scale 0-4)**	**Firmness (Newton)**	**Colour of peel (scale 1-5)**	**Colour of flesh (Chroma)**			**TSS (Brix)**	**pH**
	Local	**Export**						**L***	**a***	**b***		
Esparza	0.79	0.62	621.1	2.2	0.38	10.59	1.91	73	10	59	8.21	3.32
Garabito	0.76	0.67	551.5	2.6	0.48	9.09	1.99	70	14	64	9.21	3.54
Orotina	0.73	0.66	516.8	2.2	0.61	7.99	1.54	70	8	57	8.30	3.54
Puntarenas	0.78	0.69	536.7	2.4	0.29	8.50	1.92	74	10	57	8.80	3.65
San Mateo	0.79	0.73	508.2	1.9	0.49	2.86	3.29	66	14	65	10.62	3.68

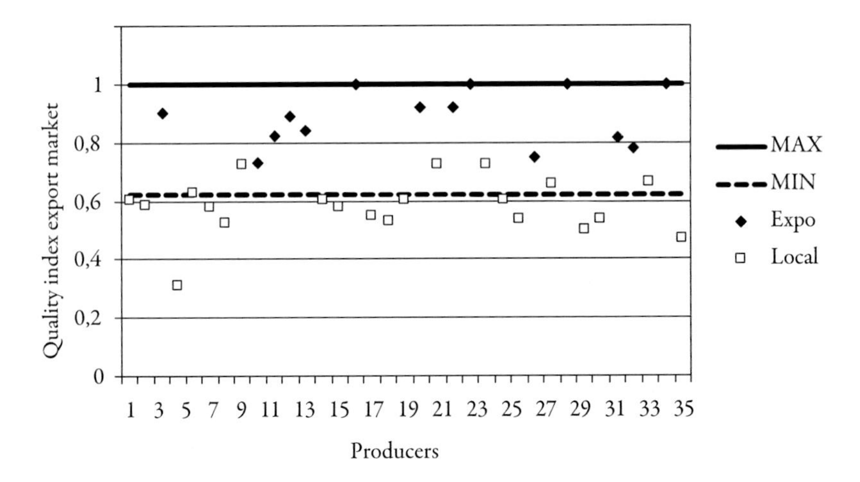

Figure 3.5. Quality scores of producers at the export market.

We performed a t-test analysis and corroborated that local and export quality indices are significantly different in terms of quality (see Appendix B). The test shows a significant difference between the quality index for the local market and the export market, where the quality index for the export market is higher than for the local market. Tests for robustness were performed using models: to execute the basic method, three cases were randomly left out

Box 3.1.

To obtain a differentiated group between the produce for export and for the local market, we computed the following equation:

1. QiEM = 0.47*damage + 0.31*appearance + 0.16*maturity + 0.06*variability; because market quality is different, the fruit is either accepted or rejected, depending on whether it satisfies the requirements. For this reason, we changed the values for damage and appearance to a dichotomous variable (0, 1): 0 if the standard is not satisfied, and 1 if it is. A consumer panel and the judgement of experts (importers and exporters) were used to determine the weighting factors for the separate attributes within this group.
2. If value for damage is greater than or equal to 0.5, then 1 is used, otherwise 0; if the value for appearance is greater or equal to 0.5, then 1 is used, otherwise, 0.
3. Because of a pre-selection process on the farm before sending the mangoes to the packer and market segmentation due to differentiated quality standards in the two markets, we lowered the value of the index for the local market by 25% for the damage and 50% for the appearance. Therefore, the final functions are as follows:

   If the mangoes were for the local market, the damage is 75% of 0.47; if the mangoes were for the export market, the damage is 100% of 0.47.

   If the mangoes were for the local market, the value for appearance is 50% of 0.31; and if the mangoes were for the export market, the value for appearance is 100% of 0.31.

   Export market producers QiEM = 0.47*damage + 0.31*appearance + 0.16*maturity + 0.06*variability.

   Local market producers QiLM = 0.399*damage + 0.15*appearance + 0.16*maturity + 0.06*variability.

and the analysis run three more times. All the analyses have shown that the models are robust (see Appendix C).

3.6.3 Farm-household characteristics

We now can relate the quality performance index for each producer to the set of farm- household characteristics (see Table 3.2). The estimated model for quality performance explains 70% of the data variance. Mango quality is related negatively to the producer's bargaining power and input costs. Positive relationships toward quality are found with producer's age, mango area and family labour. Hence, older producers operating a large mango area, with high family labour use, low investments in inputs, and low bargaining power tend to deliver higher quality mangoes. It appears that producers that are able to develop an economy of scale for their production system due to their resource endowment will deliver better quality. One option for the mix of the resources is to increase family labour use, reduce inputs and increase mango area. Nevertheless, we have no data regarding the actual decision making of the producers with respect to the argument of economies of scale.

Table 3.2. Farm-household characteristics influencing quality performance.

Quality index		
Variable	**Coefficient**	**Significance**
Age	0.480	0.003***
Experience	-0.198	0.231
Mango area	0.619	0.001***
Bargaining power	-0.650	0.001***
Hired labour	-0.110	0.575
Family labour	0.306	0.035**
Input cost	-0.463	0.003***

* significant at 10%, ** significant at 5%, *** significant at 1%.
Adj. R2 = 0.696.

3.6.4 Price characteristics

Table 3.3 shows the relationships between price characteristics and quality performance, both dry and rainy season prices are significant for quality performance but the trend of the coefficient is inverse. Mango price is defined by the supply and demand mechanism. Mango production has two seasons: dry and rainy. The dry season is: 1) the natural season for mango production; 2) the Costa Rican export window; and 3) according to the data, the best season for higher prices and higher quality mangoes. The rainy season is: 1) not the natural season for mangoes; 2) the period when the mangoes are sold mainly in the local market, because they do not comply with the minimum quality requirements for export; and 3) a time when price fluctuates. If there are huge quantities of mangoes on the market, the buyers will be selective,

Table 3.3. Price characteristics influencing quality performance.

Quality index		
Variable	**Coefficient**	**Significance**
Rainy season price	-0.515	0.001***
Dry season price	0.261	0.051*

* significant at 10%, ** significant at 5%, *** significant at 1%.
Adj. R2 = 0.324.

and control the quality by buying only the high quality mangoes and refusing low quality mangoes. If quantities are not plentiful, buyers will accept any kind of mango, regardless of the quality, and pay higher prices for lower quality mangoes.

3.6.5 Behavioural characteristics

We have sought to identify the effect of behavioural characteristics on quality performance in order to identify relevant attributes of the contractual relationship between producers and buyers that could influence quality characteristics. Key attributes of contract choice refer to important factors in reaching an agreement, attributes that influence product management, determining factors for selecting an outlet, the relationship with the buyer and bargaining power attributes. Factor analysis was performed to reduce the number of variables in the analysis (see Appendix D).

Table 3.4 presents the results of the analysis where quality performance is related to these contractual attributes (for a better understanding of the construction of the factors used, see Appendix E). We have divided the table into five sections: 1) important for reaching an agreement, 2) relationship with buyers, 3) determinants of outlet choice, 4) attributes influencing product management, and 5) bargaining power attributes.

Among the attributes that were included in *important to reach an agreement*, we found that traditional contracts and friendship have a positive influence on quality performance. Producers relying on friendship and traditional agreements (hand-shake and verbal agreements) will have higher quality mangoes. The *relationship with buyers* has both positive and negative influences on quality performance. A long-term relationship has a positive effect on quality. This is in line with the above description. Mango producers from Costa Rica are looking for long-term relationships with buyers. Uncertainty, information cost, negotiation cost, and looking for markets are some of the transaction costs affecting this kind of relationship. Trust and loyalty influence quality performance negatively; hence producers that are not trustful and loyal will deliver better quality. It seems that although producers are looking for long-term relationships, they are still not completely certain about their partners' business intentions and strategies. It is relatively common in the agro-export sector of Costa Rica to find that producers do not receive the complete payment for their produce at the end of the season because of a contract breach with the buyers.

In relation to the *determinants of outlet choice* attributes, we found positive and negative influences on quality performance. Friendship and a good reputation are positively related to quality, so producers might decide where to sell depending on the reputation of the buyer and the market. They are trying to reduce uncertainty, and thus increase the information cost of searching and screening. To get better prices faster, justice and fairness as determinants of outlet choice affect quality performance negatively. Among the *attributes influencing product management*, the hold-up problem has a positive effect on quality performance; hence

Table 3.4. Relationships between quality performance and contractual attributes.

Quality index	
Variable	**Coefficient**
Important for reaching an agreement	
Traditional contract and friendship	0.326***
Amount of control by the buyer	++
Outcome (characteristics)	0.146
Relationship with buyer	
Trust and loyalty	-0.274**
Long-term relationships	0.460***
Informative and strict	++
Determinants of outlet choice	
Justice and fairness	-0.909***
Friendship and good reputation	0.634***
Better price, fast payment	-0.361***
Hard to bargain with	++
Attributes influencing product management	
Technical package inputs and training	0.099
Friendship, information exchange	0.070
Hold-up	0.628***
Bargaining power attributes	
Risk taker	-0.426***
Commitment tactics	++
External options	-0.523***
Wealth	++

* significant at 10%, ** significant at 5%, *** significant at 1%.
++ Not in the analysis.
Adj. R2 = 0.883.

verticalisation and asset specificity play an important role in the improvements in quality. The results showed that producers acknowledge the benefit of specialisation and the ties between producer and buyers. In terms of *bargaining power attributes,* risk-adverse producers and producers with few buyers will be able to deliver higher quality mangoes. This behaviour is related to the standardisation of production processes. It would appear that mango producers are willing to take the role of followers instead of being leaders in their sector.

3.7 Discussion and conclusion

The quality characteristics of mangoes from different producers are assessed to identify suitable quality indicators, including weight, irregularity of the shape, damage (fruit fly, brown spots, internal damage, bruises and scars), firmness, colour of peel and pulp, total soluble solids and pH. Based on the opinions and preferences of Costa Rican consumers and traders, a quality model has been proposed for the construction of a quality index consisting of four key attributes: *damage*, *maturity* (including colour of peel and pulp), *appearance* (including weight and irregularity of the shape) and *variability*, indicating the variability amongst the mangoes of one producer. The weighting factors for the individual attributes are based on feedback from consumer panels, while the group weighting has been established using expert opinion from selected traders.[16] After obtaining the descriptive statistics of quality performance, categorical regression analysis is used to relate quality performance to (objective) farm-household characteristics and (subjective) perceptions of the existing relationships between producers and traders.

We find that the effectiveness of economic incentives for quality improvement depends greatly on individual farm-household characteristics and existing marketing arrangements. Elder entrepreneurial mango farmers do not accept risk and have low bargaining power, but they are engaged in stable, long-term, trustful relationships with traders and are better able to maintain quality standards as required by the market. Technological regimes for mango production based on the use of (more reliable) family labour and sufficient access to inputs and market information lead to higher quality. Control practices by the buyer appear to be effective for improving quality.

The price mechanism was divided into two types: one for the rainy season and another for the dry season. We found that the former has a negative relationship to mango quality, and the latter has a positive effect. This reflects how mango production is affected by supply and demand, and indicates that quality standards shift with changes in the market. Mangoes are produced artificially in the rainy season; trees are forced by chemicals or damage (producers cut the first layer of tree bark so that the trees will be stressed and start production) to produce after the natural period. A producer must use many insecticides to prevent pests and damage in order to be able to harvest those mangoes

Regarding the behavioural characteristics related to important attributes for reaching an agreement, traditional contracts and friendship have been found to have a positive relationship to mango quality, indicating that mango producers prefer less control and instead rely on friendship and their reputation with loyal buyers. In a similar vein, factors influencing the choice of an appropriate buyer influence the quality in both directions. Quality will be affected negatively if buyers are just and fair, and also when the buyer offers a better price and fast

[16] Since the producers are sampled at the end of the dry season, this could lead to a certain under-assessment of quality because at that time the quality of the mangoes tends to be somewhat lower.

payments, but it will be affected positively if they have a long-term relationship and a good reputation. Producers see the fact that some buyers offer better prices and cash payments as a temporary device for one season, and consider these buyers rather opportunistic and not good for business. Instead, most mango producers are looking for long-term relationships, given the perennial nature of the crop.

Bargaining power also has a negative impact on quality. If the producers tend to be risk-takers and have several outlet choices, the mango quality delivered will be lower, because the degree of verticalisation in the transaction and business relationship will be lower. Relationships in the mango business of Costa Rica are based on reputation, friendship and a certain amount of specialisation to buyers; hence, the governance structure is of great importance in the chain.

The economic incentives to improve quality from the producers' perspective may be summarised as follows: 1) Friendship and long-term relationships have a positive effect on quality, so transaction costs will decrease and the selection of the business partner is shown to be of great importance for doing well in the mango sector. 2) Reputation also has a positive effect on quality; therefore, information on the buyer and seller is a key element in developing a good reputation and thus producing high quality fruit. 3) Verticalisation is another way to increase quality. Contract formation must make producers and buyers bear a certain amount of risk according to their possibilities. Vertical integration comes with a number of rules for production and includes strengthening relationships with only a few partners. 4) Better prices and fast payments are not incentives to produce better quality; consequently, producers prefer reputation, friendship and long-term relationships rather than the possibility of getting extra surplus in the short term.

This research leads to an innovative approach for relating socio-economic characteristics and behavioural perceptions of the producers to the intrinsic quality attributes of mango production as viewed by consumers. This analysis is a clear step forward in filling the gap in understanding suitable economic incentives that will make producers responsive to consumer preferences and demands. Further research is required to fine-tune the quality index and generalise the approach for quality assessment of other commodities. Additional research is also necessary for a better understanding of the relationship between quality and the different types of price and non-price incentives.

Chapter 4. Bargaining power perception and revenue distribution in the Costa Rican mango supply chain: a gaming simulation approach with local producers[17]

4.1 Introduction

The mango supply chain in Costa Rica includes many different agents involved in numerous transactions and oriented towards multiple market outlets. Mango transactions differ in terms of volume, quality, price and delivery frequency and the produce can be sold either at the local market through wholesalers or at international markets through multinational trading companies. Relationships between producers associations, (local and international) traders, retailers and consumers are structured through a complex sequence of transactions.

Contract choice in supply chains is based on a number of structural and behavioural factors (Williamson, 1998). Agency characteristics related to partnership networks, negotiation skills and wealth might influence bargaining power. In addition, the volume supplied and the quality of the produce influence the price. Behavioural factors related to trust amongst agents and perceptions regarding contract breach also influence contract choice (Muthoo, 2002). In this chapter, we are particularly interested in two issues: (1) disentangling the factors and motives underlying bargaining power, and (2) distinguishing the factors affecting the distribution of revenues. Wealth, good partnership, market imperfections and negotiating skills are considered to be particularly important in shaping bargaining power. In addition to bargaining power, variables such as risk perceptions, contract choice and bounded rationality may determine the distribution of revenues.

We developed a stylised gaming simulation called Mango Chain Game (MCG) to assess the specific attributes that influence bargaining power and the partition of revenues. The MCG includes a description of agency roles for all participants in the mango supply chain. Attributes of their transactions are recorded, permitting contract breach, hold-up and repeated contracts. Random uncertainty is introduced in production and trade to account for unknown external factors (i.e. climatic hazards or logistic risk). The MCG can be played with groups of 7-13 participants. Here we describe the results of five sessions with real mango producers from Costa Rica. The results of the sessions include a set of single or repeated transactions of mangoes with

[17] An adapted version of this paper (written with Sebastian Meijer) was submitted to the International Journal of Chain and Network Science. An earlier version has been published in Bijman, J., Omta, S.W.F., Trienekens, J.H., Wijnands, J.H.M. and Wubben, E.F.M (eds) (2006). International agri-food chains and networks. Management and organisation. Wageningen Academic Publishers, The Netherlands.

specified price, volume and quality attributes. In addition, all of the participants answered a detailed questionnaire at the end of the game to assess their bargaining power.

One important aspect of supply chain transactions is related to the distribution of bargaining power amongst agents (Nash, 1950; Rubinstein 1982). Agency choice is therefore influenced by the anticipation of the behaviour and expected response of potential trading partners. Considerations of fairness and expectations regarding trust and reciprocity are of vital importance for engaging in single or repeated transactions (Berg *et al.*, 1995). Individual perceptions of bargaining power and differences in transaction costs can thus be attributed to the place and position of agents within a network hierarchy. Consequently, we pay particular attention to an analysis of individual and transactional attributes that shape bargaining power and revenue distribution.

It is frequently argued that smallholders face major disadvantages with respect to bargaining (due to their limited supply volume) and that the distribution of revenues is dictated by more powerful downstream agents. We would like to gain insight into how negotiation skills and partnership relationships influence the governance structure used, measured via risk distribution and contract choice. Further insights into the use of governance structures could permit the identification of strategic options for improving smallholders' positions.

This research provides an analytical framework and empirical assessment of the agency characteristics and perceptions that influence bargaining power and distribution of revenues. The analysis itself covers two levels. First, the bargaining power of the participants is explained. Performance of contracts is measured in terms of revenues of the participants at the end of the session. Secondly, the outcomes are explained by relating agency characteristics with contractual attributes, using a multiple regression analysis and logistic regression. This enables us to understand the underlying structural parameters and behavioural motives that explain bargaining power and revenue distribution.

The remainder of the chapter is structured as follows. First we discuss the organisation of the mango supply chain in Costa Rica. We identify the major agents and analyse the key attributes underlying their differences in bargaining power. Then, we outline the design of the Mango Chain Game. This is followed by an analysis of the participants' bargaining power, with reference to the underlying motives for increasing the revenue of the participants in a transaction. Finally, we discuss the usefulness of our research method for analysing the mango supply chain and we derive some policy implications for improving the efficiency of the mango supply chain in Costa Rica.

4.2 The mango supply chain in Costa Rica

Costa Rica has about 1,950 mango producers of which 60% cultivate less than 5 ha, 35% cultivate between 5 and 20 ha, and 5% own more than 20 ha (Mora, pers. comm., 2004; SEPSA, 2001). The producers are organised in different ways: large and medium-sized producers are linked to international trading companies and small producers are either independent or affiliated to co-operatives or producers' associations. There are relatively few exporters; some of them are private producers able to export by their own means. Three cooperatives or producers' associations export to Europe and one multinational company is in the business as well. The local market is more complex and the number of economic agents involved in mango commercialisation is larger. There are many intermediaries, although the exact number is unknown. These actors buy and sell mango in an unofficial way, buying rejected fruit and surplus from producers and producers' associations. There is a wholesaler market, many local farmers' markets and municipal markets.

We used data from the Ministry of Agriculture and main sources *in situ* to describe the structure of the mango supply chain in Costa Rica, distinguishing between production for the local market and the export market. The organisation of the mango supply chain is relatively simple for the export market. For the local market, the chain is more complex (Figure 4.1.) as there are a wide number of different intermediaries involved. The intermediaries play multiple roles in the chain: they buy mangoes from the producers, sell the produce to the CENADA wholesale market, or buy from CENADA to deliver to local outlets such as local retailers and the wet market.

Chain governance refers to the organisation of transactions, while the governance structure consists of a collection of rules, institutions and constraints structuring the transactions

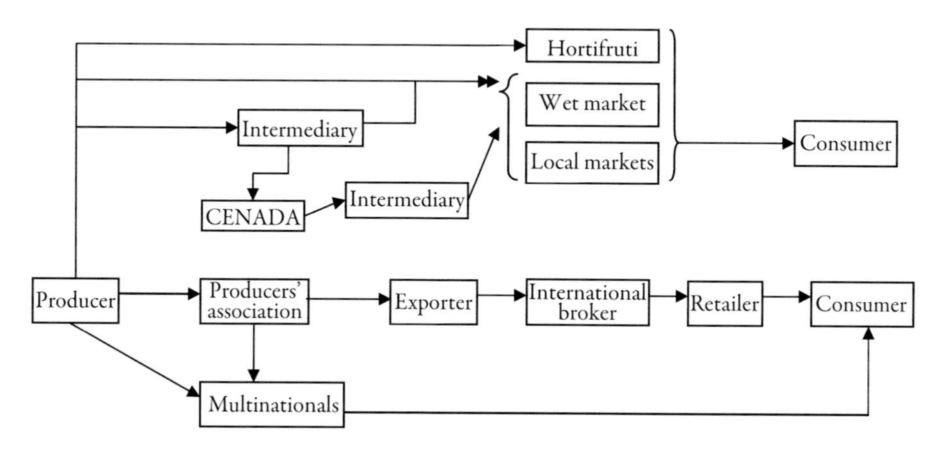

Figure 4.1. Mango supply chain in Costa Rica.

between the various stakeholders (Hendrikse, 2003). There are three governance structures: the spot market, vertical integration (hierarchy) and the network (hybrid). A governance structure affects the size of the surplus that will be generated by its effect on investment, bargaining efficiency and risk-aversion (Zingales, 1998, quoted by Hendrikse, 2003). In the case of the mango supply chain in Costa Rica, the spot market has been replaced by contractual deliveries in the export market whereas for the local market spot market, transactions are still the main governance structure. Vertical integration is present in the export market where the multinational company controls many of the actions of the producers' associations. The requirements for meeting international quality standards such as Eurep-GAP increase the presence and control of exporters and retailers on the actions of the mango producers in Costa Rica. In addition, there are some examples of network forms of governance, such as relational contracts.

4.3 Analysing bargaining power in the supply chain

The ability to secure an agreement on one's own terms is considered by Chamberlain and Kuhn (1965) as bargaining power. Chamberlain (in Leap and Grigsby, 1984) defines bargaining power as the cost of agreeing (or disagreeing) with the opponent. A party's bargaining power increases as the cost of disagreeing with an opponent decreases. In a similar vein, Slichter (1940) defined bargaining power as the cost to one agent of imposing a loss upon another agent. For Lindblom (1948), bargaining power is best defined by including all the forces that enable a buyer or a seller to set or maintain a price. The notion of bargaining power is rooted in power-dependency theory (Emerson, 1962, quoted by Yan and Gray, 2001), which states that one agent's bargaining power is derived from another's dependency. More specifically, Agent A's dependence on Agent B is directly proportional to A's motivational investment in goals mediated by B, and is inversely proportional to the availability of those goals to A outside of the A-B relation. Therefore, an agent who can develop or gain bargaining power is able to reduce dependency on other agents (Cook, 1977; Bacharach and Lawler, 1984; quoted by Yan and Gray, 2001).

A bargaining situation is a setting where individuals engage in mutually beneficial exchange but maintain conflicting interests over the terms of the trade (Muthoo, 1999, 2000). In simple words, in a bargaining situation two agents have a common interest in cooperating, but face conflicting interests over exactly how to cooperate. Muthoo (2000) explains bargaining as any process by which the agents try to reach an agreement. This process is typically time-consuming and requires agents to make offers and counter-offers to one another. Muthoo (2000, 2002) identifies seven factors that determine the bargaining power of any particular party: 1) impatience, where an agent's bargaining power is greater if someone is more patient relative to other negotiators; 2) risk of breakdown, reducing the bargaining power under the threat of breakdown; 3) outside options that increase bargaining power if other opportunities are attractive enough; 4) inside options, referring to the payoff that an agent can receive during the bargaining process, while the parties in the negotiation are in temporary disagreement; 5) wealth, since wealthier people have more bargaining power; 6) commitment tactics that refer to increasing bargaining power due to the costs of backing out of an agreement; and 7) asymmetric information that delivers bargaining power to agents with more or better access to information.

Since it would be very laborious to obtain observable and measurable characteristics for all of the bargaining power dimensions described by Muthoo, we have chosen to use a subset for this research.[18] The major determinants of bargaining power that will be used here are: 1) wealth, 2) skills of the negotiator, 3) partnership, and 4) market imperfections. Muthoo (2002) stresses *wealth* as a key determinant for bargaining power. The agent who perceives himself as wealthier will be able to exercise more bargaining power during negotiations, whereas poor agents will not be able to exercise bargaining power. Following the same line, Fossum (1982) recognises two separate aspects of bargaining power: the power inherent in the economic positions of the parties (wealth) and the attributes and *skills of the negotiator* (individual characteristics). Lee *et al.* (1998) reviewed the literature on international joint ventures and found that their bargaining power increased with the strategic *importance of partnerships* (Yan and Gray, 1994), with resource linkages between partners (Lecraw, 1984; Kuma and Seth, 1998), and with the alternatives available to the partner firm (Yan and Gray, 1984). According to Leap and Grigsby (1986), this points to some key factors that affect bargaining power, such as availability and control of resources, potential and enacted power relationships, and absolute, relative and total power. Rubinstein and Wolinsky (1985) and Leap and Grigsby (1986) emphasise that the bargaining positions, and hence the agreement reached in any particular setting, will be affected by the *conditions prevailing in the market* (market imperfections). These include the chances that each of the negotiating parties have of meeting other partners in the event that the agreement in the current negotiations is delayed (e.g. Muthoo's outside options), including the expected length of time required to achieve any alternative transaction and the expected behaviour of alternative partners as well.

For this study, we designed an analytical model to assess the different determinants of bargaining power. The dependent variable represents the revenue distribution between buyers and sellers and the independent variables are: a) the determinants of the (perceived) bargaining power of the buyer and seller, and b) the exchange conditions related to the risk attitudes of buyer and seller, the information failures occasioning bounded rationality in decision-making, and the contract choice arrangements in exchange transactions (see Figure 4.2.).

[18] The reader should remember that the main idea behind this research is the construction of a simulation game and its applicability to a research setting. A simulation game is a simplification of reality and, therefore, only a small number of behavioural and strategic dimensions of the negotiation process amongst economic agents in the chain can be addressed.

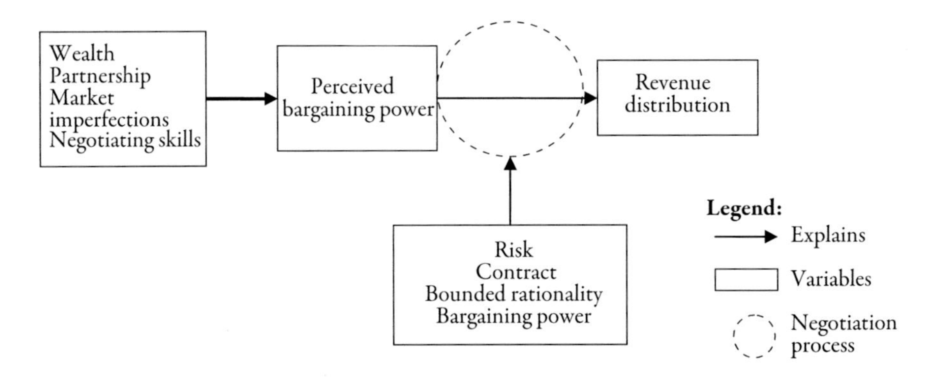

Figure 4.2. Analytical model.

4.3.1 Determinants of revenue distribution

Revenue distribution can be explained by many different characteristics. We have decided to rely on selected behavioural characteristics to explain value distribution. Hence, risk perceptions, bounded rationality (information problems), contractual arrangements and bargaining power are of key importance. Delivery agreements reached during the negotiation process provide a mechanism for sharing risk and for reducing the uncertainties produced by bounded rationality, and tend to reflect the perceived bargaining power relationships.

Close agency relationships are a vital element for supply-chain integration, but industrial experience has shown that it is difficult to contract across inter-organisational borders. Consequently, agents need to share both risks and rewards amongst members of the supply chain (Agrell *et al.*, 2002). Contracts can be considered as mechanisms to share the risk and reduce the uncertainties of the transactions. Asymmetric information in supply chains can be used by agents to reduce their own risk and costs in detriment of overall supply chain performance. Agrell *et al.* (2002) show the close interaction between risk-sharing, contracts and information asymmetries between agents in the chain. Reallocation of bargaining power may not necessarily lead to better performance and improved coordination in the supply chain. In addition, there is little or no evidence that contracts are preferred in situations where agents are more risk-averse (Roumasset, 1995).

Regarding the effect of contextual conditions on contract formation, Argyris and Liebeskind (1999) and Stinchcombe (1985) argue that the contracting agent's ability to influence the terms and conditions of contracts is highly contingent on their bargaining power (quoted by Buvik and Reve, 2002). Few buyers on the demand side will increase their bargaining power vis-à-vis the suppliers (Heide and John, 1992).

Greater bargaining power can increase stakeholders' marginal revenues, and therefore, their efforts (Chemla, 2002). Suppose that two firms hold complementary patents which could only be used together. Should they agree on a way to share revenues, they could collect a flow of revenues when the technology is used. However, unless they agree, nobody can use the patents and thus potential revenues are lost. In this context, switching costs can be arbitrarily small, since the additional fees of re-drafting an agreement would suffice (Caruana *et al.*, in press). Taking into account that the revenue is the total income produced by one agent, one can argue that the more revenue received in a specific transaction, the more bargaining power the agent will obtain, considering the income that a particular stakeholder receives from a negotiation as an expression of bargaining power (Dobbelaere, 2003).

Risk sharing is an important topic in contract design because it affects both the cost of risk-bearing and the motivation to behave in a certain desired way (Bogetoft and Olesen, 2004). According to Fafchamps (2004), who compared agricultural case studies from Madagascar, Benin and Malawi, the effect on the revenues is likely to be positive if the producer is a risk taker. Muthoo (2002) stresses that asymmetric information reduces bargaining power, and Ayala (1999) points out that insufficient information increases the uncertainty and risks of making the wrong decisions. Because asymmetric information is also part of bounded rationality (Fafchamps, 2004) agents with lower bounded rationality (lower information problems) are likely to receive higher revenue shares.

4.3.2 Negotiation

Bargaining costs are defined by Rao (2003) as the transaction cost involved in negotiations between parties to settle or bargain. They include the cost of searching, drawing up agreements, monitoring and enforcing the agreement and the opportunity cost of bargainers' time. Williamson (1975a, 1975b, 1981, and 1985) has outlined the core dimensions of a transaction: asset specificity, frequency of delivery and uncertainty. These dimensions affect the selection of different governance structures between the spot market and vertical integration. Due to bounded rationality and opportunistic behaviour, hybrid contract forms and vertical integration will ultimately replace the spot market when there are highly specific assets involved in the transaction.

In most bargaining situations, at least one player has more information about the goods traded than the other. This may lead to control over a larger share of the value added. This issue is illustrated in the marketing of 'lemons'; i.e. in the second-hand car market where the seller is assumed to have more valuable information than the buyer. Information-sharing in groups may lead to a reduction in transaction costs; likewise, it will increase trust and social cohesion and therefore the bargaining power of the partners of the group (Weinberger and Jütting, 2001).

4.4 The mango chain game

Duke and Geurts (2004) emphasise that the gaming simulation approach is relevant for strategic problem solving. It enables decision makers to analyse a multi-agent, multi-faceted real-world problem. Bringing together real-world problem owners in a well-structured and transparent gaming simulation can produce knowledge in different types of data, insights and tacit knowledge.

The effectiveness of gaming simulation for learning has been demonstrated in different settings, though it lacks a shared evaluation structure among the users of the method (Gosen and Washbush, 2004). The usefulness for learning insights into complex problems has been most prominent (Druckmann, 1997). The relevance of gaming simulation for addressing large-scale problem-solving is less documented. Authors who document some cases of using the method (e.g. Duke and Geurts (2004), and Wenzler (2003)) consider the method promising.

Gaming simulation as a data-gathering tool for research is a logical extension to the core method (gaming simulation has been used mainly for training), since it does not require any adaptations to the practices of Duke and Geurts (2004). In their *27-step guide to a successful gaming simulation,* they emphasise the importance of the operationalisation of the key concepts used in a gaming simulation. Operational (and measurable) concepts are thus required. A few authors (e.g. Kuit *et al.*, 2005; Roelofs, 2000) recognise this opportunity and have analysed behavioural change with gaming simulation. The contribution of this research is to study bargaining power and revenue distribution comprehensively using the mango chain game for data gathering. By conducting several sessions with one gaming simulation, we have tried to overcome data aggregation problems, maintaining the inputs of similar mango growers as participants.

4.4.1 Research method

The empirical results shown in this research are derived from a two-stage research approach. Figure 4.3. shows the stages of the research methodology graphically.

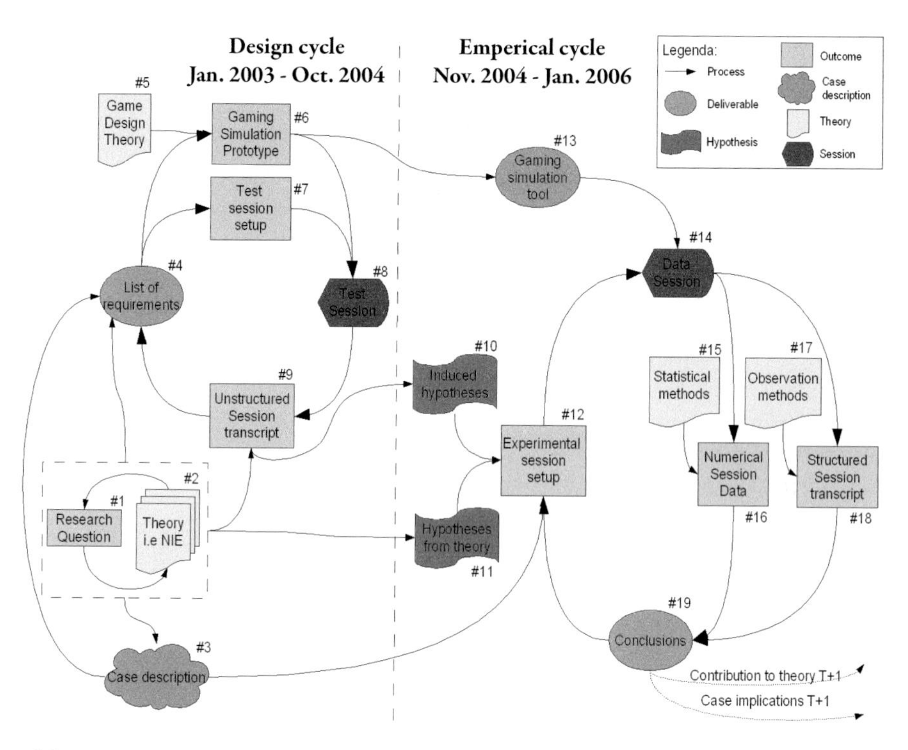

Figure 4.3. Research method with the mango chain game.

The design phase of the mango chain game (completed in October 2004) consisted of a pilot development and testing of the gaming simulation in The Netherlands and its fine-tuning in Costa Rica. Sterrenburg and Zuniga (2004) describe the Dutch development phase in detail, while Meijer, Zuniga and Sterrenburg (2005) outline the Costa Rican development and testing phase.[19]

During the empirical phase of the research method, attention has been focussed on leading data-gathering sessions using the Mango Chain Game. Based on the case description, on theory and on observations elicited from the test sessions, we constructed an experimental session set-up consisting of the load (the configuration of variables of a gaming simulation in a particular session) (see Table 4.1.A) and situation (the choice of participants, venue, measurement techniques, introduction and facilitation) (see Table 4.1.B).

[19] In these articles one can find a description of the test sessions, the fine-tuning of the simulation game, and the construction of the measurement forms. The simulation game was performed in The Netherlands several times with different participants: Dutch, Latin American, Asian and mixed groups. In Costa Rica the simulation game was tested with young professionals from different disciplines, secondary school groups, first, second, and third-year university students, university lecturers and master students. The results are consistent with our findings in this research. The results are not included in this research because our focus in this paper is on mango producers.

Table 4.1.A. Experimental session setup (Load of the game).

Topic	Configuration
Configuration of roles	3-4 producers' associations, 1-2 independent exporters, 1 multinational company (2-3 people), 1-2 retailers.
Tools per role	4 production fields per producers' association, 2 premium, 2 normal quality. One boat per independent exporter/multinational, one truck per producers' association, boats for rent by retailers.
Initial amounts (money)	Producers' associations: 200; independent exporters: 480; multinational company: 760; retailers: 730
Consumer market	Prices determined by the market; when prices go up, the supply goes down and vice versa.

Table 4.1.B. Experimental session setup (Situation).

Topic	Configuration
Participants per session	All members of one producers' association, all active in the association either as producer, as a producer's employee, or as an employee of the association itself.
Session duration	2½-3 hours of actual playing time.
Measurement of bargaining power and participant properties	Via post-session questionnaire.
Measurement of transactions	Via contract form, recorded for every transaction in the simulation, following the main three governance structures: spot market, hybrid and vertical integration.
Goal of observations	Provide qualitative additional comments to quantitative measurements.
Observation method	2 observers, plus observations from 2 game leaders.
Introduction method	Plenary explanation followed by a 'round 0': a game round that will have no consequences on the outcomes as the actions will be undone after the round ends.
Facilitation (game leaders)	2-3 controlling the behaviour of the session: one will represent the producer market (mainly production handling, natural hazards, time control); another, the consumer market (transport risk, managing the selling process with retailers); and the third, the bank's role (contract check, rent of boats, observer).

4.4.2 Description of the mango chain game

The mango chain game is played on a game board placed on a large table, with enough room for all participants to walk around. The game board reflects the structure of the mango chain (chain in Figure 4.1., board in Figure 4.5.). Participants receive game money, information sheets describing the roles for each player, products to be traded (divided into two quality categories), trucks and boats for transport, contract forms and consumer market forms. The products traded are coloured tokens representing mangoes of different qualities. The mango chain game defines four specific agents with their respective roles:

- The goal of the *producers' associations* is to earn as much money as possible. By selecting a coloured token from a covered box, they are informed about the volume of production from their producers. Producers' associations can influence the quality level of their supply by investing a fixed amount of money per production field. They are the only agents able to sell in the local market.
- The *multinational* represents a dominant agent in the mango supply chain. This is the only role that can be played by more than one person. The multinational company initially has a large amount of operating capital and its goal is to obtain as much revenue as possible. The multinational buys from producers' organisations and sells to retailers.
- *Independent exporters* are an alternative to the multinational. They have the same goal, but are considerably smaller. This becomes clear from the single person role and a more limited operational capital.
- *Retailers* have exclusive access to the consumer market. The goal of retailers is revenue maximisation. They have to deal with demand uncertainty in the consumer market.

4.4.3 Mango chain game: process description

The MCG is played in several rounds, each consisting of the following steps:

1. Production

   Production is simulated by game leaders. There are four 'production fields' per producers' association representing individual producers that belong to the association. Each of these fields produces two tons of mangoes per round in a quality determined by the quality of the production field. The amount of mango per field has a bandwidth of 50%: by picking a coloured token from a covered box, the producers' organisation has a 1 in 4 chance of 50% less production, a 1 in 2 chance of normal production, and a 1 in 4 chance of 50% more production. The coloured tokens representing mangoes are placed in a cup of the producers' association.
2. Trade

   During the trade phase, participants walk around the room and try to make contracts with one another. Contracts are signed on pre-structured forms with variables of duration, price, quantity, quality and allocation of risk. Contract choice is determined by the duration of the contract (1 round, 3 rounds, forever). Risk allocation refers to three types of uncertainties present in the gaming simulation: 1) variability in the supply from the simulated producers, 2) quality loss with a chance of 1/6 in each transport stage, and 3) uncertainty about the consumer market price. The trade phase ends with the handing in of all the signed contracts to the game leaders who will check them for accuracy and then place a coloured elastic band between the traders on the game board. The colour of the elastic band represents the duration of the contract.

3. Transportation
   During transportation of the mangoes, quality may decrease. There are three quality levels present in the gaming simulation: premium, normal and perished, each with a different price. Perished goods are destroyed immediately at no cost. For both transport modes (trucks and boats), a dice is thrown. All trucks and boats experience the same good or bad luck at once. There is a 1/6 chance that the quality will decrease by one step. Players are asked to throw the dice to check if they will have any trouble due to transport; in this way we represent real-world uncertainty such as transportation damage, storage capabilities, delays or rough handling. In a similar vein, goods that are left over after a round decrease one degree in quality, as they get older (from premium to normal and from normal to perished).
4. Contract fulfilment
   Once the sellers know how many mangoes are left after transportation, they can decide what client will get which mangoes, if there are any more contracts. In case of a mango shortage, care must be taken due to opportunities for contract breach. Payments for supplies are made at once upon delivery.
5. Consumer market
   As with the production side, consumer demand is simulated by game leaders. Retailers inform the game leaders about how much they want to sell to the consumer market. The game leader then determines prices per unit based upon the total supply. Consumers have unlimited consumption possibilities, but the price decreases when the supply grows. Separate prices are maintained for the different mango qualities. The domestic market buys all surplus fruit from producers' associations, multinational and independent exporters at a fixed low price regardless of the quality.

4.4.4 Conducting of the data sessions

Costa Rican mango producers participated as agents in all five sessions conducted. For each session, we first visited the local producers' association, to ensure in advance that the local leader of the association could mobilise enough participants. Most of the participants owned a small to medium-sized mango orchard. Some were employees of a medium-sized mango orchard or of the association itself. Roles were assigned at random to participants in a session. For the analysis of bargaining power, the perspective of local producers was of particular interest for the researchers. The roles of retailer, independent exporter and multinational roles were played from the perspective of producers. The participants enacted the roles as they perceived them in the real world.

4.5 Operationalization of the analytical model

The analysis of bargaining power and revenue distribution is based on two data sets collected during and after the game sessions. During the gaming simulation, transaction forms were used that register contract choice and the different exchange attributes (price, volume and quality). After the gaming simulation, all participants filled out a questionnaire for debriefing, with data about risk attitudes, access to information, perceptions of market access, and determinants of bargaining power. The transaction data set was comprised of 82 records (contracts) between a seller and a buyer. There were 43 questionnaires available from the debriefings. To determine the partition of the revenues both samples have been merged.

4.5.1 Bargaining power

We analyse bargaining power as influenced by wealth, partnership, negotiation skills and market imperfections. *Wealth* is related to tangible and intangibles assets in the gaming simulation. The tangible assets differed per role. Each participant was asked to give a score regarding the amount of wealth he had during the gaming simulation. Wealth is thus a self-perception of the participant in the gaming simulation. *Negotiator skills* are related to the background of the participant. For the gaming simulation we asked the participants to give a subjective measurement of their capacity to negotiate prices. *Partnership* is related to cooperation and building trust and friendship. After the gaming simulation we asked the participants to give a subjective valuation of their main business partner. Some will be friends and some will like one another in real life. The participants take these relationships with them into a session. *Market imperfections* are related to uncertainty and asymmetric information. In the gaming simulation configuration there is much uncertainty that affects agents differently. In the post-session questionnaire, we asked to what extent market imperfections influenced bargaining power.

Table 4.2. Descriptive statistics of the agency attributes.

	Producer associations		**Multinationals**		**Independent exporters**		**Retailers**	
	Mean	**S.D.**	**Mean**	**S.D.**	**Mean**	**S.D.**	**Mean**	**S.D.**
Bargaining power	7.32	2.60	8.17	1.17	6.57	1.27	5.91	1.64
Wealth	5.36	2.95	7.81	1.92	5.69	3.14	5.73	2.28
Negotiation skills	7.07	2.02	9.17	0.75	7.41	2.16	6.73	2.32
Partnership	7.86	2.50	9.00	0.89	8.06	1.50	6.73	2.76
Market imperfections	8.43	1.80	8.73	0.94	8.13	0.94	8.38	1.38

Values measured with Likert scale (from 0 to 10).

The participants of the gaming simulation had different amounts of money and individual information. Following Muthoo (2002), all variables were measured on a Likert scale ranging from 0 (low) to 10 (high). Descriptive statistics are reported in Table 4.2. We observed that multinationals had the most bargaining power, possessed the most negotiation skills and had the most wealth. Producers had relatively low wealth, but the second-highest bargaining power. This is most probably due to their hold-up capacity. Multinationals and exporters are strongly dependent on partnership to guarantee continuous sourcing. All agents are affected by market imperfections.

4.5.2 Revenue distribution

Revenue is the amount of income earned by an agent. During the sessions, participants accumulate money for three or four rounds. At the end of each session this money is counted and recorded for the analysis. Since each role had a different amount at the beginning, we used the margins above the initial amount of money as a performance indicator for each player. Figure 4.4. shows the selling/buying price and the distribution of the revenues between the participants of the sessions. In addition, we reported the margins received for the different qualities and chains that were used. To clarify the figure, let us direct our attention to the PA's box (producers' associations).

There are two prices registered here: 15 colons (the bottom price for the premium quality) and 10 colons (the bottom price for the normal quality product). PA's on average sold their product at different prices depending on the actor they faced; for example, producers received better prices when they engaged in business with the multinationals (25.5 for premium and

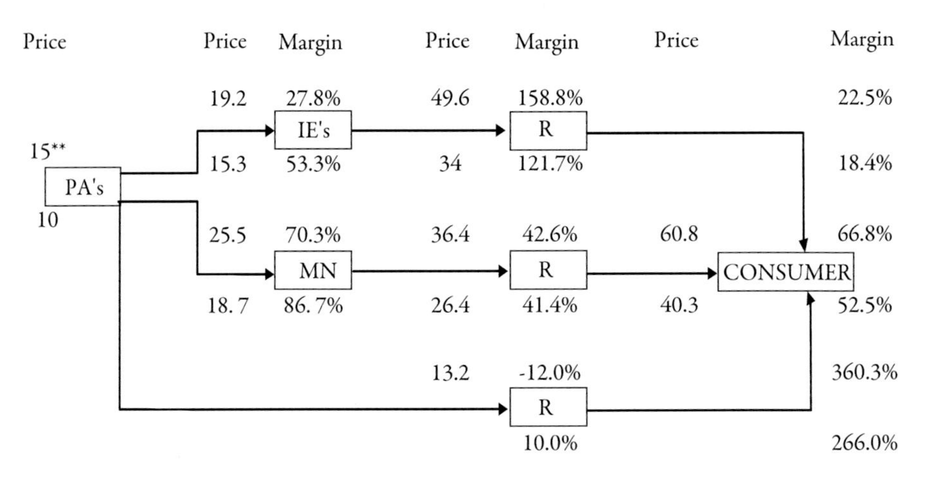

Figure 4.4. Revenue distribution in the different chain configurations.
*** Superior quality above the box and normal quality below the box. PA's (producers' associations), IE's (independent exporters), MN (multinationals), R (retailers).*

18.7 for the normal quality). When the multinational (MN) and independent exporters (IE's) sold the product to the retailers (R), the independent exporter got the best price and margin. When retailers sold to consumers, the R's got better margins when selling the product that they bought from the multinational and from the producers' associations.

4.5.3 Risk attitudes

In addition to agency attributes, all participants were asked to answer questions regarding their risk perception. We used two methods for risk assessment: 1) a direct approach where participants were asked to define their risk attitude (risk-averse, risk neutral, risk-taker) after the session; and 2) an economic experiment for risk measurement where the participants were asked to select from different lottery choices. Cross-tabulation and a Chi-square test identified a significant association between these variables. After verifying the relationship of both risk indicators, a principal component was performed in order to construct an index of risk perceptions.

4.5.4 Contract choice

The contract forms in the mango chain game permit three types of contracts: one-round, three-round and an infinite length take-over. A one-round contract is similar to a spot market mechanism: the contract specifies price, volume and quality. A three-round contract adds risk allocation and an open option. This contract requires trade partners to rely on one another as the contract will never be complete. Here the network governance mechanism comes into play. The infinite length take-over is a hierarchical mechanism where one player pays another a large amount of money to get control of his company.

Figure 4.5. shows the process of contract formation during one session. The round number is in the upper right-hand corner; each rectangle represents the game board with 4 producers' associations, 2 independent exporters, 1 multinational company, and 3 retailers. The rectangle at the left is the production market and the one at the right the consumer market; the rectangle at the top and bottom left is the local market. The triangles represent the production hazards and transportation hazards. The filled circle represents a good production season; an empty circle represents a bad one. A thick line represents a one-round contract and a thin line represents a three-round contract. As seen in round zero, most participants started to develop trade relationships and used a short-term contract. Some participants faced transportation problems (due to contract breach). One transaction had been built with a three-round contract. In round one, repeated contracts appeared (contracts with the same trade agent as the round before), and some of them were transformed into a three-round contract. In rounds two and three, it was possible to observe how some agents stuck to short-term contracts with the same traders as the round before. In effect, these contracts work like a long-term contract.

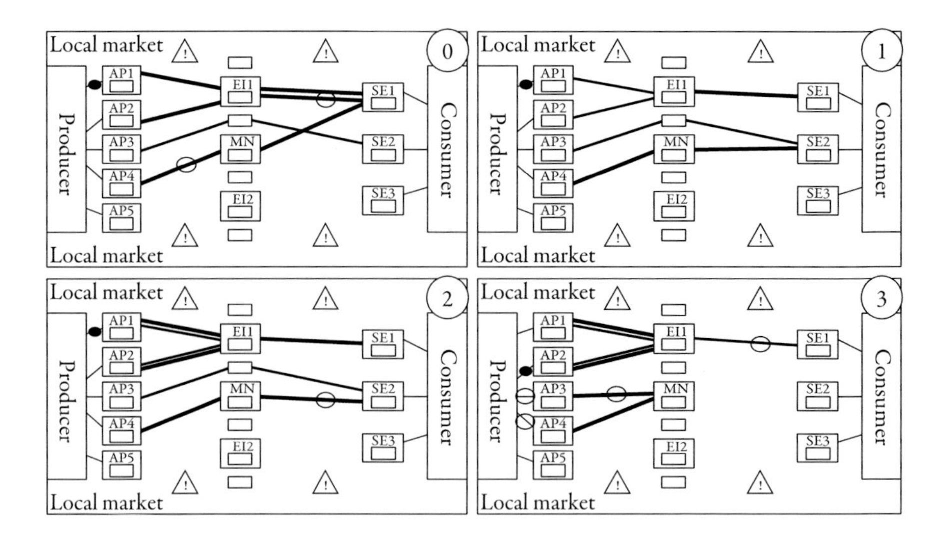

Figure 4.5. Contract configuration and construction in one session.

4.5.5 Bounded rationality

Contract making in a session takes place under partial information. Participants often stick to their trade partner from the round before, sometimes without checking other options in the market. We constructed a proxy variable for 'bounded rationality' faced by participants during the session. In the post-session questionnaire we asked all agents to identify the different types of problems faced while coming to an agreement. To be able to use the answers, we had to merge them with the information gathered from the transaction form (on the transaction form we registered contract choice and the characteristics of buyer and seller, among other attributes). Answers were classified into problems related to information and other problem areas (Table 4.3). We calculated the ratio of information problems and other problems as an index value. We used this ratio as an independent variable in subsequent regressions as a proxy for bounded rationality.

4.6 Results

The results of the gaming simulation sessions with mango producers from Costa Rica are presented below in two parts. The first part presents the outcome from bargaining power and the main influencing attributes. The second part explains the revenue distribution and how it is influenced by the attributes of our analytical model. The data analysis for operationalising the analytical model consists of three stages. Multiple regression models were estimated for the determinants of perceived bargaining power related to the agency attributes. Hereafter, we used logistic regression techniques to disentangle the differences in bargaining power between

Table 4.3. Distribution of problems faced by the participants in the gaming sessions.

	Producers' association	Multinational	Independent exporter	Retailers	Average
Information problems	0.588	0.500	0.563	0.595	0.575
Other problems	0.353	0.438	0.375	0.405	0.383
No problem	0.059	0.062	0.062	0.000	0.042

buyers and sellers, relying on the attributes of the contract partners as explanatory variables. Finally, the index of perceived bargaining power is used as an independent variable to explain the revenue distribution in the mango chain game.

4.6.1 Bargaining power

We determined how the attributes of the analytical model influence the bargaining power of the gaming simulation participants. The results are reported in Table 4.4. We observe that variables reflecting negotiation skills, partnership and wealth are all significant and have a positive effect on the bargaining power of the participants. Together these three variables explain 56 percent of the variance. Bargaining power proves to be independent of market imperfections. To strengthen bargaining power, one could increase people's skills, to motivate long-term and harmonious relationships and increase agency wealth.

Table 4.4. Determinants of bargaining power.

	Coefficient	Significance
Constant	-1.968	0.847
Market imperfections	2.958	0.811
Negotiator skills	0.325	0.017**
Partnership	0.375	0.001***
Wealth	0.194	0.055*

Note: * sign. at 10%, ** sign. at 5%, *** sign. at 1%.
Adj. R^2 = 0.558.

4.6.2 Revenue distribution between buyers and sellers

To be able to identify the factors that explain the distribution of revenues between buyers and sellers, we have constructed two different analytical models, one for the buyer's revenue and one for the seller's revenue. We disaggregate the bargaining power obtained in the first analysis for sellers and buyers depending on their contractual arrangements. The information problems related to uncertainties faced by the players during the session are also specified for both sellers and buyers. Similarly, the risk attitudes of buyers and sellers are defined using risk tests in the post-session questionnaire.

The variable *contract choice* is a binomial choice; since the participants were only able to sign one-round or three-round contracts on the contract form (the 'infinite' option was never used). The determinants of the buyer's revenue are reported in Table 4.5 and explain 47% of the variance. Factors negatively influencing the buyer's revenue are bargaining power and perceived risk of the other agent. However, neither contract choice nor the seller's perceived bounded rationality has a significant impact on revenue distribution. Consequently, with lower bargaining power and the higher risk aversion of the seller, a buyer may expect to receive higher revenues.

The determinants of the seller's revenue are reported in Table 4.6 and explain 35% of the variance. The bargaining power and risk attitude of the buyer and contract choice influence the seller's revenue negatively. The bounded rationality of the buyer is not significant. The main difference with the factors influencing the buyer's revenue is that the contract arrangement has a significant impact on the seller's revenue. Since there are only two types of contracts (one-round and three-round) in the simulation gaming, the negative sign of the coefficient means that the long-term contract decreases the revenue of the seller.

Table 4.5. Determinant of the buyer's revenue (N=82).

	Coefficient	Significance
Perceived Bargaining Power Seller (PBPS)	-0.775	0.048**
Perceived Risk Seller (RS)	-1.321	0.0001***
Perceived Bounded Rationality (PBRS)	-0.670	0.380
Contract choice	-0.317	0.108

* sign. at 10%, ** sign. at 5%, *** sign. at 1%.
Adj. R^2 = 0.466.

Table 4.6. Determinants of the seller's revenue (N=82).

	Coefficient	Significance
Perceived Bargaining Power Buyer (PBPB)	-0.379	0.039**
Perceived Risk Buyer (RB)	-2.025	0.027**
Perceived Bounded Rationality (PBR)	-1.057	0.236
Contract choice	-0.264	0.043**

* sign. at 10%, ** sign. at 5%, *** sign. at 1%.
Adj. $R^2 = 0.352$.

For both buyer and seller the perceived bargaining power of the trade partner has a negative influence on revenue. This implies that powerful traders earn more at the expense of the less powerful trade partner. This is supported by Muthoo (2002). Likewise, the risk perception of the trade partner has a negative influence on the revenue. This means that the mango chain game makes risk-averse buyers willing to pay less for the mangoes if they have to take some of the risk. Risk-averse sellers ask a higher price to be willing to take the risk. Again, this is in line with Muthoo (2002). The contract choice is only significant for the seller's revenue. As contract choice is one round or three rounds only, the seller will get higher revenues when signing one-round contracts.

4.7 Discussion and conclusion

We developed a gaming simulation approach for analysing bargaining power and revenue distribution in the mango supply chain from Costa Rica. Our approach contributes to gaming simulation methodology, as it is likely to be the first time that exchange relations are analysed in a gaming simulation setting including data that consider social relationships. Previously data gathered with gaming simulations has been mostly limited to policy making, risk attitudes, and trade games with human-to-computer interaction. In our experience, the main advantage of gaming simulation is that the results can be used to identify less tangible or unobservable issues involved in transactions, such as quality market segmentation, trust-enhancing agreements, and strategic chain configurations. The main disadvantage that we noted was that outcomes are very much dependent on rules, roles, and incentives. Therefore, an in-depth analysis of the research questions is required to determine the possible factors of influence in order to design a gaming simulation that focuses exactly on the research question and that does not limit the appearance of an explaining factor. Testing over and over again led to continuous improvement of the mango chain game. We played about four times more test sessions than data sessions.

The results from the game indicate that multinational companies obtain the highest bargaining power in the mango chain game, followed by independent exporters with a somewhat lower score, and retailers with the least power. In this sense, the gaming simulation proved to be fairly realistic in registering these differences, as confirmed by the study of Millet (2003). The producers' association in the gaming simulation shows relatively strong bargaining power compared to the information that the participants gave us during the gaming simulation sessions. This might be due to the modelling procedure that did not include the larger growers, thus leaving most room to the producers' associations in the supply market.

In the statistical analysis, buyer characteristics explain the seller's revenue and vice versa. As expected, a buyer (seller) with less bargaining power results in higher revenue for the seller (buyer). In general, stakeholders with more bargaining power will be able to take advantage of other agents. Higher risk-aversion of the buyers and/or the sellers will lead to more revenues allocated to other agents involved in the exchange relationship. In the same vein, long-term contracts in the buyer-seller relationship will lead to lower revenues (but also reduced risk) for the sellers.

The latter result is surprising, since contract choice appears to be significant only for the sellers' revenue equation. Mango producers are well aware that the type of markets in which they operate is based mainly on short-term contracts. This gives them the opportunity to remain flexible towards the changes in demand and supply that they cannot control. Producers are also trying to establish long-term relationships, but could just as well rely on repeated short-term contracts with the same partner. The latter type of contracts tends to depend more on trust or friendship.

An incentive to be taken into account in the transaction between buyers and sellers is that the latter prefer to rely on short-term contracts with the buyer that not only include price, delivery time and quality attributes, but also stipulate risk-sharing possibilities so as to redress the risk averseness of the seller to the detriment of the buyers' revenue. It is rational to believe that buyers do not want to change the status quo and are thus not interested in putting their revenue share at risk. Mango producers, however, will try to increase their bargaining power by improving their wealth, searching for good partnerships, and increasing their individual negotiation abilities. In Costa Rica, this could be achieved by offering training in negotiation skills at the National Learning Institute (INA), and the search for partners might be carried out in conjunction with International Trade Promoter (PROCOMER).

Finally, wealth appears to have a significant impact on bargaining power. In terms of tangible assets, the government could establish credit facilities for small and medium-sized producers in return for membership of a producers' organisation. This would certainly increase the producers' bargaining power, since – as observed in the gaming simulation – access to credit, resources and cooperation are key factors in the outcome of the negotiation process.

Chapter 5. Variability in quality and management in the mango supply chain: evidence from Costa Rica[20]

5.1 Introduction

Variability refers to differences within a group of individual measurements that vary from one another, either from general expectations or from historical trends. In agriculture – specifically in fruits and vegetables – variability refers particularly to *biological* variability (Schouten *et al.*, 2004), and can be described as the composite of biological properties that differentiate objects in a particular batch (Tijskens *et al.*, 2003). Therefore, commodities are disaggregated according to their main quality attributes to control for undesirable features related to size, firmness, internal and external damage, taste, colour, sugar content and nutritional value (Hertog *et al.*, 1997; Schouten *et al.*, 1997; Ketelaere *et al.*, 2006).

Variability can also be measured in terms of human behaviour and its implications for product *management*. Blackman (2001) stated that larger variability will lead to increasing monitor and control costs, and rising transaction costs as a result of supervision arrangements and possibly reduce the trust in transactions. The latter effect is mainly due to asymmetric information underlying variability in managerial terms. Berti *et al.* (1998) take production capacity and economies of scale as major indicators for quality variability, whereas Teratanavat *et al.* (2005) relate variability to firm size and fixed investments.

Variability in quality can be also measured in terms of the techno-managerial performance (i.e. Luning *et al.*, 2006). In this approach, both the technical characteristics (production and processing) and the behavioural characteristics (information and exchange) are taken into account. We follow the latter approach to analyse the determinants of mango quality variability as an interaction between biological and human processes.

Mango quality is quite variable due to a large heterogeneity in taste, flavour, aroma, colour and size, related to significant differences in managerial practices of the economic agents throughout the chain. Most consumers prefer reasonably-priced mangoes with a consistent weight, colour and consistency, and no external damage. Producers' crop management and post-harvest practices as well as delivery systems have a profound impact on observable and imminent quality characteristics (Ruben *et al.*, 2005). Although a fair amount of knowledge is available regarding the technological options for improving mango quality (Garvin, 1984; Montero and Cerdas, 2000), it is far less understood which management activities could be

[20] An earlier version of this paper by G. Zuniga-Arias, R. Ruben and T. van Boekel was submitted for publication to Agribusiness.

effective for providing actors throughout the chain with the necessary incentives for adjusting their mango production and management systems in order to enhance quality and hence cope with consumers' demands.

The organisation of the mango supply chain in Costa Rica includes a large number of different agents involved in numerous transactions and oriented towards multiple market outlets. Mango transactions differ in terms of volume, quality, price and delivery frequency, depending on whether the produce is sold at the local open market, through wholesalers, or at international markets through multinational trading companies. Relationships between associations of producers, (local and international) traders, retailers and consumers are structured through a complex sequence of delivery transactions.

This chapter presents an innovative exploratory analysis of quality and management practices throughout the mango supply chain. We point out that variability in quality depends on biological and human processes. We also suppose that there is variability in human behaviour according to the stage of the supply chain where a particular agent is involved, as well as the relations to specific market outlets. There is almost no empirical research available on these issues. A review of the literature shows several studies on specific topics, but none with an integrated approach combining natural science and social science.

The objective of this research is to describe the relationship between the variability in mango quality with differences in managerial practices amongst actors in the mango supply chain. The importance of studying the variability among actors in the chain in terms of quality and management is that gaps between local and export markets could be optimised. Reduced variability may facilitate the standardisation of procedures for meeting consumer demand. When we have a better understanding of which factors actually increase variability in mango production, they can be controlled and monitored throughout the supply chain. This chapter is structured as follows: first, variability in the mango supply chain is described, considering the relationships between quality and management within a global commodity chain perspective. Hereafter, we outline the analytical framework for an integrated analysis of technological and human factors influencing quality variability. That is followed by a discussion of our empirical results. We conclude with some implications for improving supply chain operations.

5.2 Variability and the mango supply chain in Costa Rica

The main varieties of mango grown in Costa Rica which are suitable for export are Tommy Atkins, Kent, Keith, Palmer and Smith (Central Pacific Census, MAG, 2004). In 2003, mango orchards in Costa Rica covered 8,350 ha and the quantity produced reached 35,000 tons. In only seven years (1996-2003) the area increased by 7% and the produced volume rose by 75%. This increase was mainly due to technological changes (Montero and Cerdas, 2000). Costa Rica counts around 1,950 mango producers, of which 60% cultivate less than 5 ha, 35% cultivate between 5 and 20 ha, and 5% own more than 20 ha (SEPSA, 2001). Producers are organised in different ways: large and medium-sized producers are linked to international trading companies; small producers either are not organised or are organised in co-operatives or producers associations, and their quality standards are different.

There are two main outlet choices for mango production, namely local and export with differences in quality and in handling the fruit. The organisation of the mango supply chain is relatively simple for the export market and much more complex for the local market. The local market has several sub-chains and a large number of different intermediaries involved. Quality can decline and management practices may change from actor A to actor B in the chain. For this reason, this research seeks to describe the heterogeneity of those changes in the different (current and potential) transactions and combinations of outlet choices. It also explores the main managerial factors influencing the reduction of quality variability.

5.2.1 Management practices in the mango supply chain in Costa Rica

Management practices and decisions in the orchard can affect fruit quality at the point of sale (Romano *et al.*, 2006). Management of the mango supply chain in Costa Rica is based on two international systems, the HACCP and the Eurep-GAP certifications, mainly for mangoes meant for the export market. Traditional practices (the production system developed through time by the producers) and Codex Alimentarius are used for the local market.

HACCP stands for Hazard Analysis and Critical Control Points and is defined as a systematic approach for the identification, evaluation and control of the steps in food manufacturing that are critical to product safety (Luning *et al.*, 2002). The basic objective of HACCP (1960s) is to assure the production of safe food through prevention rather than by quality inspection (Leaper, 1997). HACCP is basically designed for application in all parts of the agri-food supply chain (NACMCF, 1998). The Eurep-GAP (1997) standard is primarily designed to maintain consumer confidence in food quality and food safety. Other important goals include minimising the detrimental environmental impacts of farming operations, optimising the use of inputs and ensuring a responsible approach to worker health and safety (www.eurepgap.org). It is implemented in the entire supply chain as well. The main purposes of the Codex Alimentarius Commission (1963) are to protect consumer health and ensure fair trade practices in the food trade, and promote the coordination of all food standards work undertaken by international governmental and non-governmental organisations (FAO/WHO, 2005). The traditional practices are general standards to satisfy consumer preferences for the local market; the main sources of information include technical assistance from the government and networking with neighbours.

Just in time (JIT) and total management quality (TQM) are other international systems used in Costa Rica. Although these systems are not a requirement for accessing the export markets, they are management and production processes systems that focus on the elimination of waste, continuous improvements, employee empowerment, and data-driven decision making

(Kannan and Tan, 2002). It is hard to distinguish between JIT and TQM since both have elements in common such as consumer satisfaction and product design, thus contributing positively to manufacturing performance (i.e. Chong and Rundus, 2004; Snell and Dean, 1992; Flynn *et al.*, 1995).

To implement international standards and production systems, coordination amongst actors is needed. Supply chain management deals with the integration and coordination – both horizontally and vertically – of buyers' and sellers' decision processes with the goal of improving the fruit, and information and financial flows throughout the chain (Kannan and Tan, 2002, Gereffi, 1994). Therefore, management systems must be set in operation to coordinate the supply chain, and managerial activities will influence the final quality of the fruit. In areas where the variability of the managerial activities is low, the variability in quality will be lower as well. This does not mean that at the very moment that variability is controlled, this automatically leads to high-quality produce. The main challenge is to set a high standard for quality and then reduce the variability in quality and management, while keeping the higher quality as the main value for the complete chain.

Export-oriented producers are generally part of a producers' association. Depending on the market (USA or Europe), they must apply a different system; normally the certification process is obtained through the organisation. In the orchard producers satisfy the requirements, but they focus more on keeping the appearance of the mangoes as good as possible to reduce rejection at the packing plant. Because of the strictness of the export market, producers avoid using prohibited chemicals. Record-keeping of activities in the orchard for producers delivering to the local market is almost non-existent. Local market oriented producers obtain information on how to produce from the governmental extension officials or by networking with neighbours. Therefore, local quality management standards are not as strict as those of the export market. The producers are aware that the appearance of the fruit is one of the quality attributes valued most by the consumers.

Table 5.1 shows that the main practices present in the mango sector in Costa Rica, Codex Alimentarius is just a list of consumer needs/wishes, and it gives a small amount of specification for some of the main attributes of mango. HACCP is only a list of management prevention activities but not with performance measurements. In the case of Eurep-GAP, producers must comply with certain differentiated activities; some at 100%, others at 90% and the rest are voluntary. Asofrul (Producers' association) compliance with the international regulations is measured by practical activities with performance indicators.

5.2.2 Quality in the mango supply chain

Quality attributes valued by the different agents depend on the specific actor acquiring the product. Major actors participating in the evaluation of food quality for the export market are producers, processors, exporters, importers, wholesalers, retailers and consumers, voluntary

Table 5.1. Main features of the different quality systems from process to practice.

	Quality	Management
Codex Alimentarius	Product must be complete Practically not rotten, clean Absence of abnormal humidity, external damage, bruises, and black spots Right ripeness and appearance	Distinguishes three categories: *Extra*: superior quality, no defects *Category I*: Good quality, acceptance of the following defects, shape, skin, small bruises, sun marks, and latex marks, no more than 3, 4 and 5 cm^2 of bruises depending on the calibre *Category II*: Right ripeness, good for transport and manipulation, good appearance at market place, no more than 5, 6 and 7 cm^2 of bruises depending on the calibre and no more than 40% of the peel with necrosis Weight tolerance in quality (in grams). There are three calibres for export mango: A, B, and C. The normal weight limits are 200 to 350 gr., 351 to 550 gr., and 551 to 800 gr., respectively. There is a 10% margin for fruit outside the limits.
HACCP	Production of safe food products by prevention	Systems approach to identification of hazards, assessment of chances of occurrence of hazards during each phase (raw material procurement, manufacturing, distribution, usage of food products), and when defining the measures for hazard control. In doing so, many drawbacks prevalent in the inspection approach are avoided and HACCP overcomes shortcomings of reliance only on microbial testing.
EUREP-GAP	Food safety Quality control of the procedures and documents	Focus on activities: 1. Traceability; 2. record-keeping and internal self-inspection; 3. varieties and rootstocks; 4. site history and site management; 5. soil and substrate management; 6. fertiliser use; 7. irrigation/fertilisations; 8. crop protection; 9. harvesting; 10. produce handling; 11. waste and pollution management, recycling and re-use; 12. worker health, safety and welfare; and 13. environmental issues.
ASOFRUL	Destruction of mango samples (1 for every 10 boxes), checking internal damage, ripeness and colour	No internal and external damage, compliance with the Eurep-GAP, maturity 110 days after blooming, colour with 50% of yellow pulp, absence of fruit fly, 200-800 grams mangoes, mango should be clean and have uniform, good appearance, 7-14 calibres.

agencies and the government. Wholesalers and retailers emphasise visual attributes such as size, shape, and colour, taking into consideration consumer preferences. Customers[21] are interested in many other aspects related to food quality, such as taste, freshness, appearance, shelf life, nutritional value and food safety. Government officials are involved in health and safety regulations. Producers and processors commonly give preference to profit attributes, such as higher yields, suitability for mechanical harvesting and industrial preparation, resistance against pests and diseases, and the general appearance of the product.

The quality performance of mango is based on the external and internal quality attributes as indicated by Kader (2002). The presence of external and internal damage is a negative fruit quality attribute. Other negative external attributes include insufficient weight of the fruit, the presence of black spots, latex and other damage. The internal quality attributes are affected by the presence of mango fly, internal damage, or inadequate flesh colour, pH[22] and Brix[23] degree.

An external fruit attribute such as weight is an important fruit quality indication for the whole chain. Actors, such as producers, are paid for the kilograms of mangoes delivered to the next actor in the chain. Internal fruit attributes are essential because export markets such as that of the United States have strict laws regarding the presence of pests and disease in and on fruit (Prinsley and Tucker, 1987). Mango flesh colour relates to the maturity of the fruit. Optimal fruit maturity will be appreciated by the consumer and is a positive fruit quality attribute (Harvey, 1987; Shewfelt, 1993; Jha *et al.*, 2006). The presence of internal damage is a negative fruit quality attribute (Harvey, 1987). This damage could, for example, be due to harvest, tight fruit packing, transport or rough fruit handling in general. Further, a pH and Brix percentage measure is obtained from mango fruit juice to calculate the Brix/pH ratio. The pH of the juice indicates the acidity of the juice and the Brix percentage indicates the sugar concentration of the juice. Both attributes are important to the consumer acceptance of mango fruit (Mizrach *et al.*, 1999); the higher the ratio (Brix/pH), the sweeter the fruit and the greater the consumer acceptance.

---

[21] Customers are related to the retailers in the chain in the step before the final consumer. The final consumer has two faces in terms of quality and safety: on the one hand, consumers agree that food safety and nutritional value are very important when purchasing a product; on the other, they have a limited budget and infinite needs to satisfy. Therefore, in many cases they have to prioritise the products and the attributes of the products that they are willing to pay for (considering consumers).

[22] *pH* is a measure of the acidity and the basicity of a solution. pH values below 7 are considered acidic, whereas pH values higher than 7 are considered basic.

[23] *Brix* is used in the food industry for measuring the approximate amount of sugars in fruit juices. The higher the value of the brix measurement, the higher the sugar content.

5.3 Analytical framework

The global commodity chain (GCC) is considered as a group of networks of labour, production, distribution and marketing activities that finally result in a finished commodity (Hopkins and Wallerstein, 1986; Gibbon, 2003a; Bair *et al.*, 2006). This approach was developed by Gereffi and others within a perspective of economic policy for development (Ponte, 2002). Gereffi (1994, 1995) identified four key dimensions of GCCs: (a) the input-output structure, (b) geographical coverage, (c) the governance structure and (d) the institutional framework. We particularly focus on the governance structure where we can distinguish between producer-driven and buyer-driven supply chains and the type of governance used for coordination. Gereffi (1994) explains that buyer-driven chains are generally found in more labour-intensive sectors, where information cost, product design, advertising and advance supply management systems set the entry barriers. Most agricultural commodities tend to fall into this category (Ponte, 2002).

Consumer-driven commodity chains (in our perspective customer-driven) are changing rapidly (Muradian *et al.*, 2005) towards retailer-driven commodity chains, and coordination plays an important role in this transformation. Coordination is likely to arise in commodity chains involving suppliers in developing countries and buyers in industrialised countries (Hobbs *et al.*, 2001). Coordination is meant to ensure particular product specifications, including performance processes and logistics (Muradian *et al.*, 2005); all these requirements shift the power relationship from the producer side to the buyer side (Gibbon, 2003b).

To unravel the factors that influence the variability in mango quality, we developed an analytical framework (see Figure 5.1.) where we focus on the major causes of variability in quality and its relationship to managerial activities, as presented in our study.[24] Figure 5.1. is divided into two solid-lined figures (reflecting the technological and human management dimension, respectively) and another two dotted interfaces (reflecting managerial perceptions and outcomes). We included biological behaviour and control activities in the top solid-lined figure, related to the physical cycle of the mango. The bottom solid-lined figure refers to human behaviour and control which is influenced by information exchange and quality management activities. In the left-hand side we find the perceptions of the actors in the supply chain, while the right-hand side (dotted lines) indicates the actual outcomes of these activities.

Biological behaviour and control is composed of two main parts: (1) the perception of the behaviour and (2) the actual execution of the activities. General management intensity (GMI) and the operation index (OI) are assumed to have a positive relationship towards technological variability (TV), and subsequently technological variability (TV) is assumed to have a positive relationship with the actual value of the proxy of quality (QV). The same reasoning applies to human behaviour, where we assume a positive relationship between the quality management

[24] For further description and explanation of the procedures used for the construction of indexes for the analytical framework, see Appendix F.

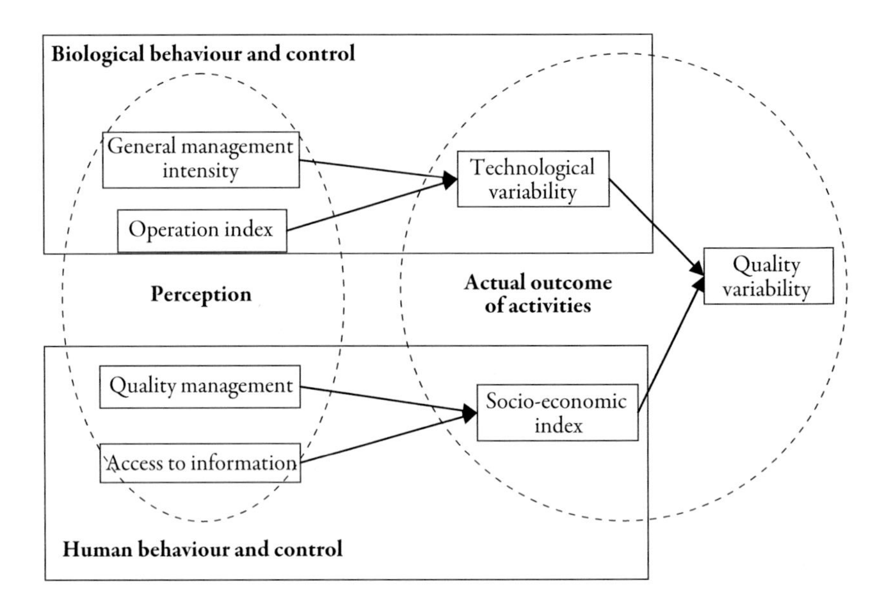

Figure 5.1. Analytical framework.

perception (QM) and access to information (AI) for the socio-economic index (SE), and a positive relationship between the socio-economic index and quality variability.

5.4 Materials and methods

The fieldwork was conducted in Costa Rica between February and April 2006. We were looking for actors in the entire chain; therefore, the areas of research were both rural and urban. We conducted 51 interviews using the socio-economic questionnaire and the mango quality analysis. The quality analysis was performed in the field. We collected products from 19 producers delivering to an export producers' association plant. Producers were located in the northern part of the country, 9 traders at the wholesaler market in the Central Valley of Costa Rica, 12 traders at the open-air farmers' market, 2 producers delivering to Hortifruti, and 9 middlemen from different parts of the country. This configuration gave us a picture of the main actors in the mango supply chain.

The sample size was not defined by any formula. We followed the snowball method (Babbie, 1992), and identified the mango producers in the production zones according to the information from our primary information sources, governmental workers and processing plant. Subsequently, other actors could be found by reference. The main reason for selecting the snowball method was a matter of uncertainty. The number of mango producers is known, but the numbers of different actors in the chain was not, nor were the relationships among them.

The main variables measured are general management intensity (GMI), quality management (QM), access to information (AI), the operation index (OI), technological variation (TV) and the socio-economic index (SE). Quality of mango is measured as a proxy of taste, which is a ratio between Brix (sugar content) and pH (acidity) (Mizrach *et al.*, 1999). For the managerial variables a Likert scale from 1 to 5 is used as indicated in Appendix F. These managerial variables are grouped and averaged to obtain an index of each particular dimension of the managerial activities. We have measured the variability within and between the groups.[25]

Many times heterogeneity and variability are used as synonyms. One may refer to the differences within a certain population, as well as to differences between groups of that population. This outcome makes it difficult and interesting for the analysis. We proposed a solution for this dilemma by using several dispersion statistics measurements such as standard deviation, standard error, coefficient of variation and meta-analysis heterogeneity measurement.[26]

Standard deviation (SD) is a measure of how well the mean represents the data. A small SD indicates that the data points are close to the mean (low variability). A large SD indicates that some data points are farther from the mean (high variability) (Field, 2002). The standard error (SE) represents the error in the mean. A large SE means that there is a lot of variability between the means of different samples. The coefficient of variation is a measurement of the ratio between the standard deviation and the average of a certain group multiplied by 100. This procedure makes it possible to compare different groups of objects from different populations in a relative way.

The term meta-analysis was introduced by Gene Glass (1976) as the analysis of the results of statistical analysis for the purposes of drawing general conclusions (Hedges, 1992). Meta-analysis aims to compare and possibly combine estimates of effects across related studies (Thompson *et al.*, 1999). In meta-analysis it is possible to determine the level of heterogeneity of a particular group based on the total heterogeneity of the population, and with the coefficient of variation we capture the heterogeneity within a group (i.e. Higgins *et al.*, 2002; Molinero, 2003a, 2003b; Raudenbush *et al.*, 1985). However, for the purposes of this research, a meta-analysis was not conducted. We have used an index to measure the heterogeneity; the formula is given in Equation 5.1 (Higgins *et al.*, 2002):

$$I = \frac{t^2}{t^2+\sigma^2} \qquad (5.1)$$

[25] We performed a discriminant analysis to try to reduce the number of groups with a similar reasoning; after the analysis, three groups were apparent: the exporters, the traders (wholesaler and middlemen), and the retailers (peasant market and Hortifruti).

[26] Only the formulas for meta-analysis variability and the regressions are presented here because the others are the standard formulas.

I indicates the proportion of variation among groups with respect to the total variability. Therefore, the proportion of the total variability attributable to heterogeneity – t^2 – is the variability among groups, and σ^2 is the variability within a particular group. If the value of I < 25%, the heterogeneity is considered low; if 25% < I < 75%, then the heterogeneity is moderate; and if I > 75%, the heterogeneity is high.

We have adjusted the index to capture the heterogeneity of a group due to the variability of the population (Equation 5.2) and the formula is as follows:

$$I = \frac{\sigma^2}{\sigma^2 + t^2} \tag{5.2}$$

We have disaggregated the scale into four clusters. If the value of I < 25%, the heterogeneity is viewed as low; if 25% < I < 50%, then the heterogeneity is medium low; if 50% < I < 75%, then the heterogeneity is medium high; and if I > 75%, the heterogeneity is high. I indicates the proportion of variation among groups with respect to the total variability, hence the proportion of the total variability attributable to heterogeneity; t^2 is the variability among groups and σ^2 is the variability within a particular group.

We have used regression analysis to investigate our assumption presented in the analytical framework. For the exploratory analysis we have run several types of regression analyses (i.e. OLS, logistic and quadratic regressions with and without a constant). OLS results are presented because they fit the data better. The functions have a specific order: first we ran the analysis for Equations 5.3 and 5.4 and then 5.5; the functions were not run simultaneously.

$$TV = \beta_1 * GMI + \beta_2 * OI + \varepsilon_i \tag{5.3}$$

$$SE = \beta_3 * QM + \beta_4 * AI + \varepsilon_u \tag{5.4}$$

$$QV = \beta_5 * TV + \beta_6 * SE + \varepsilon_o \tag{5.5}$$

Afterwards, the constructive variables of the technological and socio-economic index are used to explain quality variability.

5.5 Results

Table 5.2 shows that export market variability of the main technological and managerial variables in the study is lower than for the local market chains. Therefore, export-oriented producers are operating with higher requirements and stricter production systems than the rest of the actors in the chain. The processes related to quality and management on the retailers' side are less homogeneous than for the other markets. This may be because retailers face a larger spectrum of consumers to satisfy. The quality of the exporters turns out to be more homogeneous than that of the retailers. This supports the fact that exporters face more restrictions due to quality standards and that once the product is moving toward the consumer, quality variability increases. Regarding management, the data are not conclusive; traders seem to have more homogeneous general management intensity activities than the exporters do. One possible explanation is that traders are in the middle of the chain and face both activities in the chain (buying and selling); they must control both sides of the chain, and thus management practices (i.e. control and monitoring) must be stricter.

Table 5.2 shows the variability of the main variables in the analysis. Aside from the differences between the three stages in the chain in terms of quality, the managerial variables are less different among the stages in the chain. In this case, exporters and traders seem to have similar attitudes toward management, but not in the case of retailers that deviate from the other groups.

To determine whether the two data sets differ significantly, we have run a Kolmogorov-Smirnov test (KS test), which has the advantage of making no assumption about the distribution of data (Table 5.3). The KS test shows that there are significant differences between the attribute distributions and the different agents in the chain.

Although the distributions are similar in appearance (see Appendix G), in reality the KS test shows the opposite; the only exception is the access to information index where all three distributions are not different from one another. In relation to the quality management index, exporter and trader distributions are not different from one another. This may be because they are still far from reaching the final consumer. They have to be sure that mangoes at the consumer side will be in good condition for consumption; therefore, these actors are stricter than retailers regarding quality management. The trader and the retailers are not different in terms of the socio-economic index. As traders and retailers in the local market know one another and have done business together for some time, they may have made oral agreements and developed trust for dealing with the market information issues.

The results for quality and management presented above may have an explanation in the way each of the links in the chain faces the customer. Figure 5.2. shows that there are two well-defined customers that are different when trade occurs. On the one hand, producers have to decide whether to sell to the export market or to the local market. These markets have different standards, but both must meet the requirements of a particular buyer, which is not yet the final consumer of the produce. Thus, the produce follows the Eurep-GAP/HACCP requirements for the export market and the Codex Alimentarius and traditional production for the local market. Mangoes rejected by the export market are delivered to the local market. There did not appear to be major differences concerning management practices because all producers (for the export and wholesaler/middleman stages) are not facing the final consumer. Hence, they must have a standardised managerial system to cope with the different minimum requirements for the separate outlets that they deliver to. On the other hand, the stage that is next to the final consumer has a different managerial practice, values and quality; they will face many types of

Table 5.2. Descriptive statistics of main variables in the analysis (by actors in the chain).

	Quality (Brix/pH) **	Managerial variables*					
		General management intensity	**Quality management**	**Access to information**	**Operation index**	**Technological variability**	**Socio-economic index**
Exporter							
Mean	1.85	4.16	4.24	4.37	4.78	4.31	4.43
SD	0.17	1.06	0.59	0.48	0.34	0.43	0.19
CV	9.19	25.48	13.92	10.98	7.11	9.98	4.29
SE	0.04	0.24	0.14	0.11	0.08	0.10	0.04
Trader							
Mean	3.15	4.12	4.11	4.19	4.74	4.13	4.42
SD	0.48	0.75	0.68	0.58	0.41	0.73	0.65
CV	15.24	18.20	16.55	13.84	8.65	17.68	14.71
SE	0.12	0.18	0.16	0.14	0.10	0.18	0.16
Retailer							
Mean	3.12	3.00	3.54	4.30	4.43	3.53	4.27
SD	0.86	1.68	1.10	0.88	0.01	1.60	0.63
CV	27.56	56.00	31.07	20.47	0.23	45.33	14.75
SE	0.23	0.45	0.29	0.23	0.27	0.43	0.17
Change rate							
Ex - Tr	-	+	-	-	-	-	-
Ex - R	-	-	-	-	+	-	-
Tr - R	-	-	-	-	+	-	-

SD: standard deviation; CV: co-efficient of variation; SE: standard errors. Rate of change = (Tr - Ex) / Tr, (-) lower homogeneity of the latest link in the chain, (+) larger homogeneity in the latest link in the chain, Ex: exporter; Tr: trader; R: retailer. (Ex - Tr): relation in the chain; Tr: latest link in the chain. * For the managerial variables a Likert scale from 1 to 5 was used. **The ratio of quality is a continuous variable. Brix ($5 < x < 20$) and pH ($3 < x < 7$).

Table 5.3. Differences among actors in the supply chain. (Kolmogorov-Smirnov test).

		Exporter	Trader	Retailer
General Management Intensity				
GMI	Exporter		0.092**	
	Trader			0.002***
	Retailer	0.010***		
Quality Management				
QM	Exporter		0.722	
	Trader			0.006***
	Retailer	0.002***		
Access to Information				
AI	Exporter		0.326	
	Trader			0.375
	Retailer	0.147		
Operation Index				
OI	Exporter		0.0001***	
	Trader			0.005***
	Retailer	0.0001***		
Technological Variation Index				
TV	Exporter		0.005***	
	Trader			0.023**
	Retailer	0.001***		
Socio-Economic Index				
SE	Exporter		0.001***	
	Trader			0.578
	Retailer	0.001***		
Quality				
Q	Exporter		0.006***	
	Trader			0.039**
	Retailer	0.0001***		

* sign. at 10%, ** sign. at 5%, *** sign. at 1%.

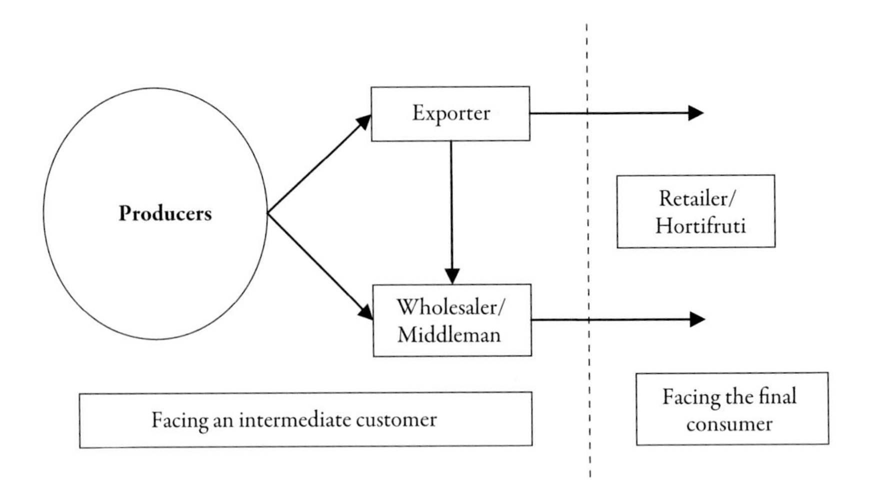

Figure 5.2. Different types of customers faced at certain stages in the chain.

consumers for different outlets. Hence, variability in the managerial practices and quality is larger for this stage of the chain.

5.5.1 Heterogeneity among the different actors in the mango supply chain

Figure 5.3. is related to the meta-analysis; it is important to note that the group with least heterogeneity is that of the exporters. The lowest overall heterogeneity is in terms of quality. Most of the managerial attributes among actors are similar and regarding heterogeneity are within the range between 25% and 75%. The exception is the quality heterogeneity of the export producers which is very low (below 25%). We performed a t-test analysis to check for the differences between actors and the main attribute value. Results for all three actors (exporter, trader and retailer) are that quality variability is significantly different, while general management intensity variability is only different for the retailer, and in the case of technological and socio-economic variation, they are different only for the exporter (see Appendix H).

5.5.2 Variability within the group

Table 5.2 showed the standard deviation for all three chain actors in the analysis. We assume that the homogeneity of the export market will be larger than for the other markets, because quality standards and international regulations for export are superior to those for the local market. The data show that our assumption is correct, at least in relative terms: homogeneity for exporters in management and quality indexes is larger than for the other actors. One exception is the general management intensity index where the trader has the lowest variability in the data. It is important to note that in the other cases variability behaves similarly. That

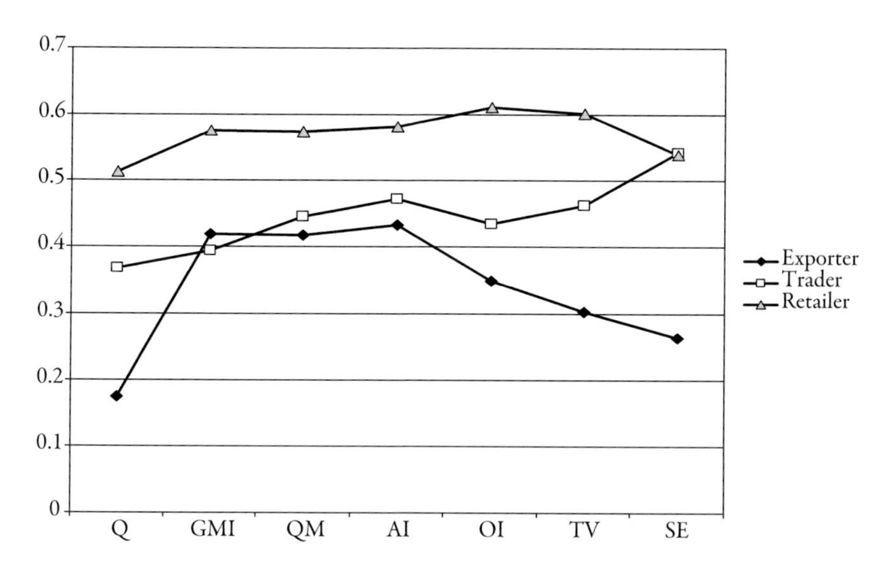

Figure 5.3. Heterogeneity among the main variables in the analysis and the actors in the chain.

means that the exporter has the lowest variability, followed by the trader and then the retailers with the highest variability.

To understand the relationships between quality variability and variability in managerial activities we ran a regression analysis. We assume managerial variability to vary positively with quality variability.

Table 5.4 shows a positive relationship between GMI and OI and technological variability, thus confirming our assumption of a positive relationship. This means that when the variability of GMI and OI increases, the variability of the technological variation increases as well. The variability of the data is explained for 44%.

Table 5.4. Determinants of the technological variability.

Variable	Coefficient	Std. error	t-Statistic	Significance
Constant	0.07	7.64	0.01	0.993
General management intensity	0.78	0.16	4.86	0.001***
Operation Index	0.52	0.25	2.08	0.043**

Note: * sign. at 10%, ** sign. at 5%, *** sign. at 1%.
Adj. R2: 0.443, Durbin-Watson stat. 2.02.

Table 5.5 shows the relationship between QM and AI and the Socio-Economic Index. It is positive as expected; therefore, when the variability of the quality management and the access to information increases, the variability in the socio-economic index increases as well. The variability of the data is explained for 12%.

We have assumed that the variability in quality of mango is related positively to technological variability and the socio-economic index. In Table 5.6 our analysis shows that these assumptions are correct. The variance of the data is explained by 32%. The socio-economic index has more weight in the relationship, meaning that in this particular case in order to lower the variability in quality of mango, the actors must work harder to reduce the differences in human behaviour and the control area of our analytical framework.

It is reasonable to believe that management activities are an intrinsic force driving the control and monitoring of technological variability. Management could have an influence in two different ways: first, the technological process (quality procedures and production processes); and second, the business relationship (contract choice, trust, volume, price, governance structure in the relationship). Information and access to it are of great importance to be able to manage business relationships.

Table 5.5. Determinants of the socio-economic index.

Variable	Coefficient	Std. error	t-Statistic	Significance
Constant	4.48	2.04	2.19	0.034**
Quality management	0.13	0.08	1.79	0.08*
Access to information	0.17	0.10	1.80	0.08*

* sign. at 10%, ** sign. at 5%, *** sign. at 1%.
Adj. R2: 0.122, Durbin-Watson stat. 1.83.

Table 5.6. Determinants of the quality variability.

Variable	Coefficient	Std. error	t-Statistic	Significance
Socio-economic index	3.75	1.42	2.64	0.011**
Technological variability	2.56	0.46	5.50	0.001***
SE*TV (interaction effect)	-0.09	0.04	-2.63	0.012**

* sign. at 10%, ** sign. at 5%, *** sign. at 1%.
Adj. R2: 0.323, Durbin-Watson stat. 2.63.

Along these lines we want to investigate which original variables used for the construction of the indexes (TV and SE) have an influence on quality variability. Table 5.7 shows the results for the variables affecting the technological variation index. The model actually explains 45% of the variance and has a Durbin-Watson statistic of 2.36. Variables that are significant and positive are input utilisation and harvest time. Hence, if the variability in input use and harvest time increases, then the variability in mango quality will also increase. In order to offer uniform quality of mangoes, producers for both markets must agree on the consumer preferences that they will satisfy.

In Table 5.8, we have analysed the relationship between the variability of variables which constitute the socio-economic index and quality variability. The model explains 17.5% of the variance and has a Durbin-Watson statistic of 1.77. Market knowledge is the only significant variable and has a positive sign; thus, an increase in market knowledge will lead to an increase in the variability of quality. This seems to be an odd result, but it is not. Imagine a producer with a limited knowledge of the market (only producing mango for a certain market); the quality that he

Table 5.7. Technological variables affecting the variability of mango quality.

Variable	Coefficient	Std. error	t-statistic	Significance
Genetic diversity	0.310	0.207	1.494	0.142
Quality samples	-0.016	0.390	-0.042	0.967
Agro- env. cond.+	-0.455	0.654	-0.695	0.490
Input utilisation	1.592	0.728	2.188	0.034**
Harvesting time	0.628	0.338	1.862	0.069*

+ Agro-Environmental Conditions; * sign. at 10%, ** sign. at 5%, *** sign. at 1%.
Adj. R2: 0.448, Durbin-Watson stat. 2.36.

Table 5.8. Socio-economic variables affecting the variability of mango quality.

Variable	Coefficient	Std. error	t-statistic	Significance
Having an agreement	-0.051	0.711	-0.072	0.942
Having a defined price	0.240	0.370	0.650	0.519
Market knowledge	5.170	1.798	2.876	0.006***
Trust in partner	0.080	0.401	0.199	0.843
Long relationship	0.137	1.631	0.084	0.934

* sign. at 10%, ** sign. at 5%, *** sign. at 1%.
Adj. R2: 0.175, Durbin-Watson stat. 1.770.

produces must be homogeneous for that market. When the same producer has more knowledge about the market (therefore increasing the variability of the knowledge/information), he can start producing mango for several channels and quality will be more varied.

5.6 Discussion and conclusions

This analysis has shown that quality variability is lower on the export side of the chain than in the local-market chain, in the case of mangoes from Costa Rica. The variability in quality increases the closer the fruit gets to the customer. This might be because actors near the customer face larger consumer segments and preferences. Producers delivering to the export market face international regulations forcing them to have a certain type of produce to meet the strict requirements for the export market. This is in line with Shewfelt (2006) who affirms that consumers do not behave uniformly, being influenced by their cultural, historical, religious, demographic, economical and social background.

Management practices are rather similar among actors in the mango supply chain in Costa Rica. This research has shown that quality variability is related to technical variables such as the input use and harvest time, both of which can be controlled and executed with precision. There is also a positive relationship towards the socio-economic variation in terms of market knowledge. Producers or economic agents that know the market better are able to deliver the preferred mango quality for certain types of customers. This is also related to management and access to information. It is important to observe that price knowledge and prior agreements do not have an influence on quality variability.

Both biological and human behaviour need particular managerial activities to produce high quality fruit. Thus, further research should be conducted along these lines, to be capable of disentangling the different management activities related to biological and human behaviour and to develop accurate performance indicators.

This research suggests that there is a gap between export and local markets, mainly in terms of quality requirements rather than in regard to managerial practices. Hence, the market outlet defines the product characteristics valued by the consumers. The research conducted also points to the positive relationship between technological and socio-economic variability and the variability in quality; therefore, the greater the variability in management the greater the variability in quality.

Further research will be required to explain consumer preferences concerning variability in quality and technical managerial decisions throughout the chain. In this vein, Romano *et al.* (2006) conclude that quality has different meanings for different stakeholders (producers, distributors, consumers, etc.) and consumer acceptance seems to be the most important factor to be considered. It would also be meaningful to extend the analysis to the role of the international regulations and their effect on the export and local market homogenisation.

Chapter 6. Coordination and governance in the mango supply chain from Costa Rica

6.1 Research issues

This research has been conducted in Costa Rica based on the analysis of mango supply chains for export and for the local market. By differentiating between these two market channels, it is assumed that distinct behaviour and performance could be expected related to the coordination and management principles in force in each segment. Attention has been focussed on some of the most critical variables that refer to supply chain coordination and governance, market choice, quality management, contracting practices and bargaining procedures, and management practices throughout the chain.

We have focused on the quality of the fruit and its relationship with chain management throughout the chain because this provides the framework for economic incentives towards agency coordination within the chain. Within this general framework, several other related issues emerged and have been operationalized, depending on the particular purpose of each chapter. The main topics of research that are addressed refer to market selection, economic incentives to improve quality, bargaining power and revenue distribution, quality variability and management variability throughout the chain.

6.1.1 Research questions

The main research questions addressed in this study can be recapitulated as follows:
1. What are the determinants of market outlet choice for local and export market producers?
2. What kind of price and non-price attributes can be used to provide incentives for higher quality mango production?
3. How can bargaining power in the mango supply chain be assessed and what factors determine revenue distribution?
4. How does management variability affect quality variability in mangoes, and what managerial factors will contribute to quality homogenisation?

We will first summarize the most important findings derived from the research and then draw some conclusions with respect to the opportunities and constraints for enhancing mango supply performance through improved procedures for agency coordination.

6.1.2 Market outlet choice

We started with an analysis of the factors that influence market differentiation between the export market and the local market in the mango sector in Costa Rica, and identified the main determinants for differences in outlet choice. We therefore focus on the importance of production systems organisation, farm household characteristics, price attributes, and market context for outlet choice decisions.

Surprisingly we find that less-experienced and younger mango producers with a risk-adverse attitude tend to be more export-oriented. Trust in the buyer does not play a major role in the selection of the market outlets in the case of mango in Costa Rica. Some of these findings are not fully in line with other research, but may be explained by the fact that the latter type of producers are more likely to accept the strict contracting conditions of exporting firms, where full alliance with the delivery requirements provides a degree of certainty. Moreover, prices are not a significant variable for outlet choice, most likely because mango producers in Costa Rica are price-takers. Our findings with regard to the production system are in line with the literature; we found a positive relationship between the orientation towards the export market and the mango area and mango production. The market context results are also in line with the theory: the higher the degree of complexity and the formalities in the purchase condition and the closer the production site is located to the market, the higher the probability of delivering to the export market.

6.1.3 Economic incentives for quality improvements

We analysed different economic incentives for enhancing quality performance in mango deliveries. These economic incentives can be divided into price attributes and non-price attributes. We also constructed a quality index based on the measurement of several objective (external and internal) mango attributes. Hereafter, we made an exploratory interdisciplinary (beta-gamma) analysis of the interlinkages between both aspects. Our results indicate that farm-household characteristics – such as age, area of mango orchards, bargaining power, family labour, and input cost – have a decisive influence on quality. In general, older producers, with large mango plantations, using much family labour, reducing their input cost and with low bargaining power will be able to produce higher quality mangoes. Moreover, contractual conditions for quality incentives can serve as a basis for developing friendships; a relationship with the buyer should be based on trust and the need to strive for long-term relationships. To select appropriate outlets, producers are mostly looking for reputable and friendly partners. Hold-up is positively related to quality; therefore, producers will be willing to be less flexible while having a closer relationship with mango buyers. In line with the hold-up problem, for delivering high quality mangoes, producers should be risk-averse and have a low number of buyers.

6.1.4 Bargaining power and revenue distribution

After sorting out the factors that determine the bargaining power of economic agents in a transaction, the revenue distribution between buyers and sellers can be explained. Bargaining power is positively related to negotiating skills, partnerships and wealth. Our findings confirm Muthoo's (2002) constructions of bargaining power, except for the impact of market imperfections. Market imperfections include the chances that each of the negotiating parties have for engaging with other partners in the event that the agreement in the current negotiations is delayed, considering the expected length of time required to achieve any alternative transaction and the expected behaviour of alternative partners as well. In the mango supply chain in Costa Rica, such market imperfections do not play a major role in determining the bargaining power of the economic agents in the chain.

Any transaction has at least two agents involved, a buyer and a seller, both seeking cooperation, but disagreeing on how to cooperate. Revenue distribution is the performance indicator of the negotiation process. Because of the dichotomy of transactions, revenue distribution has been divided into two parts: that of the buyer and that of the seller. The buyer's revenue will be larger when the bargaining power of the sellers is low and when the seller is risk-adverse. The seller's revenue will be larger when bargaining power of the buyer is low and the buyer is risk-adverse. In addition, the seller's revenues will be larger when signing spot market contracts (short-term and low dependency with buyer's production system). Consequently, expectations about the behaviour of the other partners involved in the transaction are critical for explaining the type of contractual arrangements and the resulting revenue distribution.

6.1.5 Quality and management variability throughout the chain

We conducted an exploratory analysis to find out whether objective quality attributes can be related to specific managerial practices of different actors in the supply chain. Its main objective is to explore the relationship between the variability in mango quality due to differences in managerial practices amongst actors in the mango supply chain. The importance of understanding the variability among actors in the chain in terms of quality and management is that market differences between local actors and exporters could be optimised. Reduced variability may facilitate the standardisation of procedures for meeting consumer demands. We find that the closer the economic agent is to the final consumer, the greater the quality variability. This is particularly observed at the local market, where retailers face a wide diversity of consumers. As expected, for the export market, quality variability is substantially lower. Producers have to follow strict international delivery requirements (HACCP and Eurep-GAP) to access these markets.

Regarding management practices, there is greatest homogeneity in the export market, followed by the traders, and least for the retailers. In general, traders maintain more homogeneous management compared to other actors in the chain, because they have to control activities at both ends of the supply chain (buying and selling). Technological and socio-economic variations have a positive effect on quality. It is empirically confirmed that quality variability is most dependent on the variability of management, related to both the biological system and human behaviour. Input utilisation and harvesting time are technological activities that most affect quality variability positively, while in the socio-economic realm, quality variability is positively affected by market knowledge.

6.2. Supply chain coordination and governance for quality

Nowadays it is no secret that supply chain management is of great importance for doing business. Gereffi (1994) showed the key role played by management coordinating actions throughout the chain. Coordination in the chain will lead to different governance structures (Williamson, 1981 and 1985), which are dependent on the presence of transaction cost, asset-specific factors, and uncertainties and frequency of delivery that are specific for each commodity. Different commodities will have different transaction costs – dimensions, and values – and, therefore, different governance structures emerge for coordinating actors throughout the chain. While many actors are involved in the management of perishable products, most perishable chains are characterised by being controlled by the retailer. The retailer is the actor in the chain that is closest to the consumer and thus knows what to put on the shelves to attract as many consumers as possible. Successful supply chains require both suppliers and buyers to have the ability to valuate quality in order to avoid adverse selection problems (Gorton *et al.*, 2006).

Quality is a key motor of any chain, but it can be addressed from different perspectives. One can say that quality is only related to the physical attributes of the fruits (in this case), while others consider it as a variable related to the managerial practices and services provided between actors in the chain. We believe that quality is a composite of internal and external product characteristics (biological) and managerial practices (human behaviour). If quality is related to both biological and human attributes, monitoring and control must have procedures and protocols for both dimensions. Technological management will be suitable for controlling production and post-harvest treatment of the fruit, but incentives to induce desirable behaviour of the actors must be also available. This research has shown that the type of agreement and its features are an appropriate way to induce the required actions.

Internal and external quality indicators must be controlled by laboratory research. Most of the attributes can be assessed by simple procedures that can be performed at the packing plant. The problem with these procedures is that they are very time-consuming and the fruit must be destroyed to carry out the analysis. At the moment when the performance indicators are set, the control and monitoring of the bottlenecks in the supply chain can be addressed. The definition of a group of performance indicators brings up some other issues mentioned in our research. The definition of certain standards of quality requirements for different types of consumer makes market selection a very important decision for the actors in the chain.

Market selection, according to Root (1994) and Koch (2001), can be based upon three basic approaches: (1) selection with no particular market entry, characterised by a short-term horizon with no systematic criteria; (2) selection in accordance with an existing market entry strategy, and (3) a systematic approach considering available alternatives.

Market selection is thus based on many different aspects but the effort of the producers to deliver to the export market or the local market is of key importance. This effort is related to various structural factors:

a. *Distance to the market.* Mango producers are concentrated in two main production areas (Guanacaste and Puntarenas on the North, and Alajuela in the Central Pacific Region). Most of the producers delivering to the export market are located in the Northern part of the country, where the export producers' associations and the packing companies set up their operations.
b. *Plot and production certification.* Producers delivering to the export market must get an international certification for entering the market. These certifications, in the case of small producers, are issued by the certification body to the producer association.
c. *Homogenisation of the produce.* Homogenisation will make the producers a specialised asset for the buyer of the fruit. However, the rejection rate may increase as well due to the standardised quality parameters, motivating the producer to look for an alternative outlet for the rejected fruit.
d. *Quality variability.* Quality variability is greater in the local market; therefore, managerial and quality practices are less strict compared to the export market. Local market producers have a lower rejection rate, and local consumers have a wider range of mango varieties to choose from.

If controls and monitoring can help to reduce the variability of mangoes between the local and export market, then standardisation of products and procedures can be achieved. The main problem is that retailers in the export market are most concerned with having available standard-quality produce in reasonable amounts throughout the year. Local market retailers know that consumers expect to have mangoes at different stages of maturity, as well as with a variety of sizes, colours, aromas, and appearances. The small size of the local market makes it quite hard to widen export market regulations into a rule for the local market. However, the local market may seek international certifications to improve management practices and quality and safety standards. Instead of homogenisation, the export market might also want to offer a wider variety of mangoes at different stages of maturity. This option will basically depend on the existence of market niches that are willing to pay for this type of product. The local market can accept the standardisation of procedures for mango production and new technology, but not the option of selling only one type of mango.

Another aspect of supply chain governance refers to the exchange and delivery relationships between agents. Bargaining power is related to the producers' ability to negotiate with the buyers. The economic agents' bargaining power depends on the market choice they make.

Producers choosing the export market enter into a more vertical relationship with the buyers, and the verticalisation of this relationship makes the livelihood strategies of small producers less flexible. Local producers, on the other hand, are more flexible when using the spot market as main outlet for trade. Both markets are essentially different in terms of the governance mechanism they apply, and – as expected – many producers want to sell their produce in the spot market, while the buyers prefer to enter into more hybrid forms of contractual deliveries, where the producer must accept the buyer's rules for production and processing.

Bargaining power will be greater for small producers if they are able to increase their wealth and negotiating skills. These two features of economic agents in the chain cannot be modified in the short run. Mango producers in Costa Rica are price-takers, and thus they are not able to influence prices at international markets. Throughout our research, prices have not proved to be significantly important for mango transactions at any level. In perishable product chains, the power shifts quickly to the retailer, and this is also the case for the mango chain. Producers need to sell their fruit fast; otherwise they will lose it. The physical appearance of the product is essential for both local and export consumers. Information asymmetries between farmers and processors lead to most market failure (Gorton *et al.*, 2006), and thus induce differences in bargaining power throughout the chain.

6.2.1 Market choice, market failures and chain governance

A very important decision at any level and for any commodity refers to the choice of appropriate market outlets. In the case of the mango supply chain in Costa Rica there are several options. Producers might choose between local and export markets. For the export market, they have two main options (US and EU markets), and for the local market they have at least five options (wholesaler, peasant market, middlemen, a private company and municipal markets). To decide where to sell, the agents must collect information about prices, volumes, costs, distance to markets, and quality requirements, among others.

Market failures appear when the supply and demand mechanisms do not work properly. In a world of bounded rationality, market failures are a constant factor in most decisions. Therefore, the information problems, the gaps in access to information and the constrained ability to make productive use of this information, often causes huge performance differences between actors in the supply chain. The differentiation between agents allows some of the actors to be better able to deal with these asymmetric information problems.

Coordination of governance of the supply chain is one way to tackle this information problem. Following the institutional paradigm, there are three main governance mechanisms: spot markets, hybrid and vertical integration, all of which have their advantages and disadvantages (Williamson, 1985). Concerning the spot market, in the presence of information problems, when asymmetric information situations take the lead in business relationships, small mango producers tend to avoid using this mechanism. Mango producers in Costa Rica are mostly risk-adverse; they prefer to sell their mango at the gate of the orchard, concentrating on production rather then engaging in marketing and commercialisation. Verticalisation has been studied several times and findings show that some scholars believe it will help smallholder producers, while others have demonstrated quite the opposite (i.e. Key and Runsten, 1999). In our research, verticalisation is important when producers become involved in the export-oriented chain, or when they are part of a chain with the supplier of fruits and vegetables of major retail companies of Costa Rica. For other outlet choices, producers tend to avoid vertical integration. We showed, however, that vertical integration or hybrid forms of governance that are closely related to vertical integration are most suitable for improving the socio-economic performance of producers, thus reducing the gaps between the local market and the export market and enhancing the overall performance of the supply chain.

6.2.2 Power, revenues, and coordination

The performance of the chain can be also be addressed in terms of the quality of the produce with respect to satisfying consumer preferences. Due to governance problems in any specific chain, bargaining power and revenue distribution are of great concern to the economic agents. Perishable product chains are retailer-driven, where most power over transactions is in the hands of the buyers and not at the producers side. Power is a key determinant of revenue distribution. An agent capable of enforcing power can influence the equity distribution among actors in the chain. The quality of the products demanded is strongly determined by the consumer and the requirements imposed by retailers. These market signs have to be transmitted throughout the chain up to the producers.

Vertical and horizontal coordination could be helpful to reduce the differences in power and revenue distribution among actors in the chain. Vertical integration and coordination might also enhance quality and management practices throughout the chain. Horizontal coordination among producers will increase their bargaining power and hence their revenue share. Horizontal coordination also enables them to produce a larger volume and comply with market requirements for export and local markets. It has been proven that coordination among producers enhances their strategic position in the chain (Saenz-Segura, 2006). To produce high-quality products, producers look for horizontal coordination and larger volume for increasing their bargaining power, while they require a vertical integration mechanism to be able to produce the mango that consumers at local and export markets ask for. The possibility of having two agreements (one horizontal and one vertical) makes this strategy difficult to accomplish. Producers should organise a firm for horizontal coordination purposes, where the middleman will be the agent involved in linking different consumer preferences with production. This middleman might maintain a vertical relationship with the producers and a hybrid relationship with the other actors in the chain. In order to perform well, the new middleman should be able to get information from the other actors in the chain and the consumer preferences in both markets. Quality would then be defined by the consumers, and delivery agreements (such those used in Scandinavia, the US and The Netherlands; see

Bogetoft and Olesen, 2004) based on quality produce incentives are required for adequately dovetailing these demands with producers' practices.

6.3. Methodological innovation

We relied in this research on a combination of several methodological approaches. We used field surveys, laboratory analyses for objective measurements of quality criteria, and simulation gaming. The surveys were performed in the standard way. The main innovation is that we combined survey results with objective measurements of quality attributes (based on QFD from a natural science perspective) and with the socio-economic characteristic of different actors in the chain (from a social science perspective). As a result, an empirical application of a practical interdisciplinary research approach could be developed.

The simulation game has been used previously for educational purposes in this field; but we have developed here a simulation game for research purposes. In the simulation game, players take the role of different actors in the chain, receive information about their role and then play this role taking all their business experience with them into the simulation. Consequently, the participants in the simulation game behave as they would in reality. We used a wide variety of tools to gather relevant information regarding the decisions taken in the game: observations of the game sessions, photographs of the round table sessions, contract forms, debriefing and questionnaires.

An additional lesson that we have learnt from the interdisciplinary approach is that social scientists are able to understand the complex behaviour of producers (in this particular case) more comprehensively, since they rely on more than just socio-economic variables for the day-to-day decision-making. By being able to grasp this complexity and include some of these motives in the analysis of certain situations, the researcher is (in a more holistic way) better able to understand the reality in which the producers are immersed. In the case of interdisciplinary research with social sciences, simulation games have several advantages over the traditional econometric analysis. They enable the researcher to observe the behaviour of real actors in the chain, and thus reveals how (and why) transactions take place. For this purpose we developed some forms and procedures to collect the information systematically during the simulation game simulation and use this data subsequently for further statistical analyses. It was also possible to have a discussion with the participants at the end of the session in a debriefing. In addition, since the interaction between the researcher and the actors in the chain takes place in a relaxed setting, better feedback is obtained. Simulation games have been commonly used in this field for training, but here we are proposing that it be used for research purposes as well.

6.4. Policy implications

The focus of this research was on smallholder producers that are at a 'disadvantage' with respect to other actors in the chain. Mango producers in Cost Rica are price-takers; they cannot control the mango supply worldwide or even regionally, and therefore non-price attributes appear to be more important than price attributes.

Governance of the chain is essential for a good performance by the producers involved. For this reason, institutional agreements have been studied systematically. Most producers in the mango chain in Costa Rica prefer to sign short-term contracts with buyers looking for better prices the following season, while some producers establish contractual relationships with buyers. As an end-result of these strategies, short-term contracts easily become repeated contracts, thus building long-term relationships. During all of the interviews, questionnaires and observations in orchards and at different outlets, the behaviour of the agents was quite similar and consistent: first they try a short-term relationship, and then they use it as a basis for establishing a long(er)-term relationship.

Our results show that (good) partnership has a strong positive influence on the selection of long-term contracts. In other words, stronger and durable partnerships between buyers and sellers allow for more complex and complete contractual conditions. Other variables influencing contract selection are bargaining power and information problems. Individually, actors that have high bargaining power will choose the spot market governance structure, whereas actors with lower bargaining power will prefer hybrid forms of governance structures. Transactors facing significant information problems tend to prefer complex hybrid contracts in the negotiations and those with fewer information problems will look for spot market governance structures.

In negotiations on a commodity, buyers and sellers meet and bargain over certain attributes; when bargaining is successful, a specific type of agreement is reached. For such agreement, the sellers will prefer the spot market for the governance structure, whereas the buyers will choose hybrid arrangements. Consequently, the coexistence of contracts is likely to prevail unless market information failures are reduced.

Transactions amongst agents may include some other characteristics, in addition to the length of the contract, such as procedures for dealing with hold-up problems, contract breach and repeated contracts. Agents in the mango chain are well aware of the possibility of a contract breach. Normally, the contract formation process consists of several offers and counter-offers before anyone is able to reach an agreement. In this process, several offers from different buyer and sellers are posted to the same agent, exposing them to a simultaneous prisoner dilemma (as reflected in the simulation game).

Contract breach and hold-up problems are very common in the mango chain. Producers face the problem that buyers set the rules of the game, usually because they have most bargaining power in the relationship. Small producers facing a contract breach are incapable of enforcing the contract. Consequently, other actors may confront hold-up problems as well. Producers face the problem of specialising in mangoes in general or in mangoes for the export market.

Their usual strategy is to sell produce in many different outlets and to maintain several other crops as well, thus avoiding specialisation. Producers' associations and cooperatives also face problems with equipment to sort fruit and cooling systems to maintain the quality of the mangoes. Their reaction to the hold-up is to purchase machinery capable of sorting several kinds of fruits and to use the cooling system for other products as well. What can be done then to improve business? Producers and producers' associations may try to increase their bargaining power – and thus become better capable of enforcing their contracts – by taking several actions, some in the short term and others with a long-term perspective.

In the short term, producers could improve coordination, transforming their small-scale production system into a common production organisation by means of agency cooperation. They could sign a single contract with the buyer, and rely on third party control for enforcement. This also provides other advantages for complying with larger delivery volumes to buyers. In addition, this would enable negotiations on better conditions for the transaction (eventually relying on hold-ups in the bargaining process).

In the long term, producers and producers' associations could try to increase their wealth by additional training but in areas other than just production, for example improving their\ capacities in negotiation tools, management systems, marketing, processing, and financial strategies. In addition, they may also get better credit as a group, thus increasing the tangible part of their wealth.

These collective action strategies do affect the relationships within a group of producers and the producers' associations because the control and monitoring costs will increase so as to keep the group united. In this situation, opportunistic behaviour and moral hazards may arise and affect the cooperation. Problems of contract enforcement, trust and reputation need to be balanced against the possibilities for establishing long-term relationships (something that many producers value highly).

When incentives throughout the chain and between the transacting actors in the chain are aligned, transparency and information flow might become available to all the actors involved in a contract. If actors get a 'fair' piece of the cake, clear performance indicators must be applied for both vertical and horizontal relationships. Therefore, a sound management system for information exchange and registering cash and produce flows throughout the chain will be of great value. The final structure of the chain could be one where each actor receives a share of the entire operation and is considered as a partner of the other actors (stakeholders) in a particular chain

We know from experience that quality and management are related concepts and represent key processes for enhancing performance of any chain, and in this the mango chain is no exception. Quality is also one of the keys for entering new markets. We explained which incentives can be used to improve quality throughout the chain, in particular at the producer level. It is equally important to create procedures for analysing consumer behaviour in potential markets, and thus have the information required to decide whether the producers' association, the producer, or the nation as a whole has the potential to penetrate those markets successfully. This type of research is very costly, but in a consumer-driven supply chain such as the mango chain, it is essential to align the incentives of every actor in the chain. The national government must become involved in this task; otherwise, the agent paying for the analysis will be the one getting the information and hence increasing his own bargaining position in the chain. Here a public-private partnership seems to be the preferred option, but in the case of Costa Rica, there is little experience with these kinds of relationships. The fact that professionals and entrepreneurs do not usually trust each other leads to the problem of how to create a good partnership between the public sector and the private sector. We suggest that (academic) professionals should present themselves as experts in research and be willing to work with the entrepreneurs to solve the day-to-day problems that they face in the business environment. Professionals can make use of contracts to be able to conduct their research (i.e. use the data for scientific publications) and to tailor this knowledge to identify solutions for the practical problems at the enterprise or sector level. As our research has shown, such a dual relationship is a process and hence must start with small efforts to build up the trust and reputation necessary for longer-term relationships.

Another strategic management option is to reduce the gap between the local market and the export market and to produce only export-quality products. This is not simple to accomplish, in the case of mangoes in Costa Rica. There are two different, well-defined customer segments when trade takes place. Producers have to decide whether to sell to the export market or to the local market. These markets maintain different standards, but both must meet the requirements of a particular buyer (trader or retailer), which is not the final consumer of the produce. Thus, the fruit complies with Eurep-GAP/HACCP requirements for the export market and the Codex Alimentarius criteria and traditional production norms for the local market. Mangoes rejected by the export market are delivered to the local market. It is likely that there are no major differences in management practices because all the producers (for the export and wholesaler/middleman stages) do not face the final consumer; therefore, they must have a standardised managerial system to cope with the different minimum requirements for separate outlets where they deliver. On the other end, the stage that is closest to the final consumer does maintain different managerial practice values and quality standards, since they have to face many types of consumers at different outlets. For this reason, variability in managerial practices and quality at this final stage is much greater compared to all the other stages in the chain. How can the same management and quality requirements be implemented for both (local and export) chains? Some efforts have been made by one local company that delivers fruits and vegetables to the main supermarket chain. It has created a new private standard for quality and management of fruits and vegetables, but their norms are still at the level of international standards, even though the products still vary according to the type of clients.

A final major issue is the large number of different actors involved in the mango chain for the local market. From a transaction cost perspective, it would be better if we could reduce the number of actors required to reach the consumer. In particular, the role of local middlemen (those that buy surplus mangoes from the producers, and also those buying at the wholesale market to sell the product on the street or at the farmers' markets) remains fairly ambivalent. If the agents involved in the mango supply chain can be identified and if the actual demand for mangoes is known, market forces might be put to work and the invisible hand would make the rules and regulations to work. The role of the government could then be restricted to control and monitoring activities.

6.5. Further research

For a better understanding of the decision-making processes of the different actors in the supply chain, detailed information would be required on the production function for different market outlets. It would also be essential to determine the market and marketing mix of the producers to understand their multiple market orientation. Both aspects could provide further information on the backwards implications of market outlet choice for the strategic behaviour of mango producers. Consumer preferences and characteristics deserve further analysis, considering also potential markets for other commodities.

Recent approaches to market selection scarcely consider the importance of the local market, and are mostly applied to companies entering foreign markets. The analytical tradition is to focus on competitive advantages of countries; whereas we consider firms as the principal agents to decide whether they are able to export on their own. In our research, the analysis of market selection is oriented towards smallholder firms that cannot export by themselves, and where the local market is a significant source of income. This implies that we need to consider new organisational forms for smallholders, or assess the competencies of the already existing organisations, to identify their real and potential comparative advantage. Most current analyses of market selection are related to firm/country relationships (Beim and Levesque, 2006) and relationships between large companies (Gorton *et al.*, 2006). However, buyer-seller networks in developing economies remain relatively understudied (Andrabi *et al.*, 2006). The importance of the local markets has generally been underestimated, and greater attention should be given to reducing the quality gaps between local and export market regimes.

Regarding the options for improving quality throughout the chain, a further fine-tuning of the quality index is required to generalise the approach to the quality assessment of other commodities. Further research is also necessary for a better understanding of the relationship between quality and different types of price and non-price incentives. It would also be important to understand how the creation of value added throughout the chain, and its relationship to revenue distribution and equity objectives, can be influenced in an indirect way by focussing on quality upgrading in particular stages of the supply chain.

Until now, there have only been a few studies that describe the relationship between smallholders and large companies or processors (i.e. Key and Runsten, 1999). Nevertheless, the options for contract enforcement still remain unclear; smallholders will easily break the contract whenever they receive better offers from other buyers. Relational contracts and self-enforcing contracts appear to be useful mechanisms for strengthening the buyer-seller relationships. Different contract specifications can be used as incentives to promote mutual relationships, simultaneously improving the quality of the produce and reducing opportunistic behaviour.

Further research must also be conducted to explain consumer preferences with respect to the variability in quality and the related technical managerial decisions throughout the chain. In this vein, Romano *et al.* (2006) conclude that quality still has quite different meanings to different stakeholders (producers, distributors, consumers, etc). Consumer acceptance seems to be the most outstanding factor to be considered. It would therefore be useful to broaden the analysis towards the role of the international regulations and their effect on the export and local market homogenisation. Some additional analyses of bargaining power and governance in the chain can be conducted, particularly considering their relationship with quality. The mango chain game may be generalised and applied to different types of market settings.

It would be worthwhile carrying out an in-depth study on the effect of the shifting of power balance toward retailers and its effect at the producers' level, focussing on the implications for quality variability, diversity of produce and private international standards. To tackle these problems, new organisational forms for producers and processors can be studied to explore the idea of a national brand for Costa Rican commodities to counteract the emerging power concentration at the retailer side.

Some general transaction attributes require further research. To align the incentives within the chain, options for cooperation amongst actors in the supply chain deserve to be studied in depth. The role of trust and reputation in quality management and institutional arrangements of the chain are considered essential. Given the need to operationalise these concepts, it will be necessary to address opportunistic behaviour, free riding and moral hazards from a principal-agent perspective.

Regarding the methodological approach, it is considered highly fruitful to continue the development of interdisciplinary (beta-gamma) research in different supply chain settings. This approach certainly provides a more comprehensive understanding of reality. Gaming simulation could be applied for particular research questions and perhaps the emerging experimental economics framework will help increase our insight into the reactions of the different supply chain actors, to risk and uncertainty. This implies, however, that far more attention must be given to the development of innovative and participative procedures for data collection and analysis.

Finally, it is essential for research and the development of science to rely on interdisciplinary approaches for a better understanding of complex realities, and to gain more insight into the strategic business behaviour of the actors in the chain. Therefore, new methodological tools are to be developed that permit an adequate operationalization of this interdisciplinary analytical framework. In addition, research needs to expand from a (comparative) static analysis of supply chains toward a more dynamic analysis, taking into account the interactions over time and space, and considering feedback between quality management, information exchange and market outlet choice. This will certainly provide a worthwhile challenge for future projects and new research.

References

Agarwal, R. and M. Gort, 2002. Firm and product life cycles and firm survival. American Economic Review 92 (2), 184-190.
Agarwal, S. and S. Ramaswami, 1992. Choice of foreign market entry mode: impact of ownership, location and internalization factors. Journal of International Business Studies 23 (1), 1-27.
Agrell, P., R. Lindroth, and A. Norman, 2002. Risk, information and incentives in telecom supply chains. International Journal of Production Economics 90, 1-16.
Aichian, A., 1950. Uncertainty, evolution and economic theory. Journal of Political Economy 58, 211-221.
Alexander, C., R.E. Goodhue, and G.C. Rausser, 2000. Do quality incentives matter?. Department of agricultural and Resource Economics, University of California Davis. Working Paper No. 00-029.
Ali, M., 1995. Quantifying the socio-economic determinants of sustainable crop production: an application to wheat cultivation in the Tarai of Nepal. Agricultural Economics 14, 45-60.
Amir, R., I. Evtigneev, T. Hens and K. Schenk-Hoppé, 2005. Market selection and survival of investment strategies. Journal of Mathematical Economics 41, 105-122.
Andersen, O. and A. Buvik, 2002. Firms' internationalization and alternative approaches to the international customer/market selection. International Business Review 11, 347-363.
Argyris, N. and J. Liebeskind, 1999. Contractual commitments, bargaining power, and governance inseparability: Incorporating history into transaction cost theory. Academy of Management Review 24, 49-63.
Ayala, J., 1999. Instituciones y Economía. Una introducción al neoinstitucionalismo económico. Fondo de Cultura Económica, México.
Babbie, E., 1992. The practice of social research, Wadsworth, Belmont CA.
Bacharach, S. and E. Lawler, 1984. Bargaining: power, tactics, and outcomes. Jossey-Bass, San Francisco, C.A.
Bain, J., 1951. Relation of profit rate to industry concentration. Quarterly Journal of Economics 65 (3), 293-324.
Bair, J. and E. Dussel, 2006. Global commodity chain and endogenous growth: export dynamism and development in Mexico and Honduras. World Development 34 (2), 203-221.
Baker, W., 1990. Market networks and corporate behavior. American Journal of Sociology 96, 589-625.
Baker, W., 1994. Networking smart. New York: McGraw-Hill.
Batt, P. and N. Parining, 2000. Price-quality relationship in the fresh produce industry in Bali. International Food and Agribusiness Management Review 3, 177-187.
Ben-Porath, Y., 1980. The F-connection: families, friends, and firms and the organization of exchange. Population and development Review 6, 1-30.
Berg, J., J. Dickhaut and K. McCabe, 1995. Trust, reciprocity and social history,
Games and Economic Behavior 10, 122-142.

Berti, P. and W. Leonard, 1998. Demographic and socioeconomic. Determinants of variation in food and nutrient intake in an Andean community. American Journal of Physical Anthropology 105, 407-417.

Blackman, A., 2001. Why don't lenders finance high-return technological change in developing-country agriculture? American Journal of Agricultural Economics 83 (4), 1024-1035.

Boger, S., 2001. Quality and contractual choice: a transaction cost approach to the polish hog market. European Review of Agricultural Economics 28 (3), 241-261.

Bogetoft, P. and H. Olesen, 2004. Design of production contracts. Lessons from theory and agriculture. Copenhagen Business School Press. Copenhagen.

Brewer, P., 2001. International market selection: developing a model from Australian case studies. International Business Review 10, 155-174.

Buvik, A. and T. Reve, 2002. Inter-firm governance and structural power in industrial relationships: the moderating effect of bargaining power on the contractual safeguarding of specific assets. Scandinavian Journal of Management 18, 261-284.

Buzano, P., 1997. Fases históricas de la producción del mango en Costa Rica. Maestría en Desarrollo Integrado de Regiones Bajo Riesgo, Mimeo. 8 p.

Caruana, G., L. Einav and D. Quint, forthcoming. Multilateral bargaining with concession costs. Journal of Economic Theory (article in press)

Chamberlain, N. and J. Kuhn, 1965. Collective Bargaining, 2nd ed. New York: McGraw-Hill.

Chemla, G., 2005. Hold-up, stakeholders and takeover threats. Journal of Financial Intermediation 14, 376-397.

Chetty, S. and D. Holm, 2000. Internationalization of small to medium-sized manufacturing firms: a network approach. International Business Review 9, 77-93.

Chiarelli, C., R. Dieci and L. Gardini, 2002. Speculative behavior and complex asset price dynamics: a global analysis. Journal of Economic Behavior and Organization 49, 173-197.

Chong, V. and M. Rundus, 2004. Total quality management, market competition and organizational performance. The British Accounting Review 36, 155-172.

Christensen, P.R., 1991. The small and medium-sized exporters' squeeze: empirical evidence and model reflections. Entrepreneurship and regional development, 3, 49-65.

Codex Alimentarius, 1993. Norma Mundial del Codex para el mango. CODEL Stan. 184 - 993.

Coleman, J., 1988. Social capital in the creation of human capital. American Journal of Sociology 94: 95-120.

Consejo Nacional de Producción (CNP), 2006. Mango análisis del mercado. Boletín 2, año 12, Septiembre. San José, Costa Rica

Cook, K., 1977. Exchange and power in networks of international relationships. The Sociological Quarterly 18, 62-82.

Crosby, P.B., 1979. Quality is Free. McGraw Hill, New York.

Dicken, P., 1986. Global Shift, Industrial Change in a Turbulent World. Harper & Row, London.

Dobbelaere, S., 2003. Estimation of price-cost margins and union bargaining power for Belgian manufacturing. International Journal of Industrial Organization 22 (10), 1381-1398.

Driscoll, A. and S. Paliwoda, 1997. Dimensionalizing international market entry mode choice. Journal of Marketing Management 13, 57-87.

Druckman, D., 1994. The educational effectiveness of interactive games. In: Crookall, D. and; Arai, K. (Eds.) Simulation and gaming across disciplines and cultures, SAGE publications.

Duke, R. and J. Geurts, 2004. Policy Games for Strategic Management. Pathways into the Unknown. Dutch University Press, Amsterdam, The Netherlands.

Dunning, J.H., 1981. International Production and the Multinational Enterprise. Georgellen & Unwin, London.

Dunning, J., 1988. The electric paradigm of international production: a restatement and some possible extensions. Journal of International Business Studies 19, 1-31.

Edgington, D., 1987. Influences on the Location and Behaviour of Transnational Corporations: Some Examples Taken from Japanese Investment in Australia Geoforum 18 (4), 343-359.

Engelbart, R.B., W.G.A. Frank and L.W. Rijswijk, 2001. Plantania. Business Case description. ACC. KLICT Project 'Verduurzaming ketenkennis'. Eng Dnk 2001389. August 2001, http://www.ak-acc.org/frame.html.

Ellis, P. and A. Pecotich, 2001. Social Factors influencing Export Initiation in Small and Medium-Size Enterprises. Journal of Marketing Research 38 (February), 119-130.

Emerson, R., 1962. Power-dependence relationships. American Sociological Review 27, 31-41.

Enke, S., 1951. On maximizing profit: a distinction between Chamberlain and Robinson. The American Economic Review 41, 566-578.

Erramilli, M. and C. Rao, 1990. Choice of foreign market entry mode by service firms: role of market knowledge. Management International Review 30 (2), 135-150.

Eurep-GAP, 2001. EUREPGAP. Protocol for Fresh Fruit and Vegetables. http://www.eurep.org

Fafchamps, M., 2004. Market institutions in sub-Saharan Africa: Theory and Evidence. The MIT press, Cambridge, Massachusetts, London, England.

Field, A., 2002. Discovering statistics using SPSS for windows. Advance techniques for the beginner. SAGE publications Ltd, London.

Flynn, B., S. Sakakibara and R. Schroeder, 1995. Relationships between JIT and TQM: practices and performance. Academy of Management Journal 38 (3), 1326-1360.

Fossum, J., 1982. Labor relations development, structure, process. Business Publications: Plano, Texas.

Galaskiewicz, J., 1985. Social organization of an urban grants economy: A study of business philanthropy and nonprofit organizations. Orlando, FL: Academic Press.

Gains, N., 1996. The repertory grid approach. Measurement of food preferences. London, Blanckie Academic & Professional. Editor.

Garvin, D.A., 1984. Product quality. An important strategic weapon. Business Horizons 27 (3), 40-43.

Glass, G., 1976. Primary, secondary, and meta-analysis of research. Educational Researcher 5 (10), 3-8.

Gereffi, G., 1994. The organization of buyer-driven global commodity chains: how US retailers shape the overseas production networks. In: G. Gereffi and M. Korzeniewicz (Eds.), Commodity chains and global capitalism. Westport: Greenwood Press.

Gereffi, G., 1995. Global production systems and third World development. In: B. Stallings (Ed.) Global change, regional response: the new international context of development. Cambridge: Cambridge University Press.

Gibbon, P., 2003a. The African growth and opportunity act and global commodity chain for clothings. World Development 31 (11), 1809-1827.

Gibbon, P., 2003b. Value-chain governance, public regulation and entry barriers in the fresh fruit and vegetable chain into the EU. Development Policy Review 21 (5-6), 615-625.

Goldsmith, P.D. and T.P. Sporleder, 1998. Analyzing foreign direct investment decisions by food and beverage firms: An empirical model of transaction theory. Canadian Journal of Agricultural Economics 46, 329-346.

Gosen, J. and J. Washbush, 2004. A review of scholarship on assessing experiential learning effectiveness. Simulation & Gaming 35 (2), 270-293.

Gow, H., D. Streeter and J. Swinnen, 2000. How private contract enforcement mechanisms can succeed where public institutions fail: the case of Juhosucor A. S. Agricultural Economics 23(3), 253-265.

Granovetter, M., 1985. Economic action and social structure: The problem of embeddedness. American Journal of Sociology 91, 481-510.

Greene, W., 2003. Econometric analysis. Prentice Hall, New Jersey.

Harvey, E.C., 1987. Maturity Indices for Quality Control and Harvest Maturity In: R.T. Prinsley and G. Tucker (Eds.) Mangoes a Review. The Commonwealth Secretariat, London, 39-53.

Hart, O. and B. Holstrom, 1987. 'The theory of contracts'. In: Bewley, F. (ed.), Advances in economic theory. Cambridge University Press, Cambridge.

Higgins, J. and S. Thompson, 2002. Quantifying heterogeneity in meta-analysis. Statistics in Medicine 21, 1539-1558.

Heide, J. and G. John, 1992. Do norms matter in marketing relationships? Journal of Marketing 56, 32-44.

Hendrikse, G. and W. Bijman, 2002. On the emergence of growers' associations: self-selection versus market power, ERIM Report Series research in Management, Rotterdam.

Hendrikse, G., 2003. Governance in Chain and Networks: A Research Agenda. ERIM Report series Research in Management, Rotterdam.

Hertog, M.L.A.T.M., L.M.M. Tijskens and P.S. Pak, 1997. The effects of temperature and senescence on the accumulation of reducing sugar during storage of potato (*Solanum tuberosum* L.) tubers: a mathematical model. Postharvest Biology and Technology 10, 67-79.

Hoang, B.P., 1998. A causal study of relationships between firm characteristics, international marketing strategies, and export performance, Management International Review 38 (1), 73-93.

Hobbs, J.E., 1996. A transaction cost approach to supply chain management. Supply Chain Management 1 (2), 15-27.

Hobbs, J.E. and L.M. Young, 2000. Close vertical co-ordination in agri-food supply chains: a conceptual framework and some preliminary evidence. Supply Chain Management: An International Journal 5 (3), 131-142.

Hobbs, J.E. and L.M. Young, 2001. Vertical linkages in agri-food supply chains in Canada and the United States. Canada: Strategic policy Branch for Agriculture and Agri-food.

Hopkins, T.K. and I. Wallerstein, 1986. Commodity chains in the world economy prior to 1800. Review 10 (1), 157-170.

Humphrey, J. and A. Oetero, 2000. Strategies for diversification and adding value to food exports: a value chain approach. Geneva, UNCTAD.

Huang, C.L., 1996. Consumer preferences and attitudes towards organically grown produce. Review of Agricultural Economics 23, 331-342.

Inderst, R., 2002. Contract design and bargaining power. Economics Letters 74, 171-176.

Jang, S., A. Morrison and J. O'Leary, 2002. Benefit segmentation of Japanese pleasure travelers to the USA and Canada: selecting target markets based on the profitability and risk of individual market segments. Tourism Management 23, 367-378.

Jha, S.N., A.R.P. Kingsly and S. Chopra, 2006. Physical and mechanical properties of mango during growth and storage for determination of maturity. Journal of Food Engineering 72, 73-76.

Jirón, L., 1995. 'Opciones al uso de insecticidas en el mango'. In: Opciones al uso unilateral de plaguicidas en Costa Rica. EUNED. San José.

Johnson, G.I, J.L. Sharp, D.L. Milne and S.A. Oosthuyse, 1997. Postharvest Technology and Quarantine Treatments. Ed by R.E. Litz in The Mango. Botany, Production and Uses. CAB INTERNATIONAL. 447 - 507

Julian, J.W., G.H. Sullivan and G.E. Sanchez, 2000. Future market development issues impacting Central America's nontraditional agricultural export sector. American Journal of Agricultural Economics 82, 1177 - 1183.

Juran, J.M., 1990. Juran on leadership for Quality. The Free Press, New York.

Just, R.E. and D. Zilberman, 1983. Stochastic structure, farm size, and technology adoption in developing agriculture. Oxford Economic Papers 35 (2), 307-328.

Kagel, J. and A. Roth, 1995. The Handbook of experimental economics. Introduction by Albin Roth. Princeton University Press, Princeton, New Jersey.

Kahn, M.A., 1981. Evaluation of food selection patterns and preferences. CRC Critical Reviews in Food Science and Nutrition October, 129-153.

Kay, J., 1993. The foundations of Corporate success. New York: Oxford University Press.

Key, N. and D. Runsten, 1999. Contract farming, smallholders, and rural development in Latin America: The organization of agroprocessing firms and the scale of outgrower Production. World Development, 27 (2), 381-401.

Koch, A.J., 2001. Factors influencing market and entry mode selection: developing the MEMS model. Marketing intelligence and Planning 19 (5), 351-361.

Kochan, T., 1980. Collective bargaining and industrial relations; from theory to policy and practice. Homewood, Illinois, III. Richard D. Irwin.

Kortbech-Olesen, R., 1997. World trade in processed tropical fruits. Geneva, UNCTAD.

Kotler, P., 1999. Marketing management: Analysis, planning, implementation and control (10th ed.). Englewood Cliffs, NJ: Pretence-Hall, Inc.

Kuit, M., I. Mayer and M. de Jong, 2005. The INFRASTRATEGO game: an evaluation of strategic behaviour and regulatory regimes in a liberalizing electricity market. Simulation & Gaming 36 (1), 58-74.

Kumar, S. and A. Seth, 1998. The design of coordination and control mechanisms for managing joint venture-parent relationships. Strategic Management Journal 19, 579-599.

Lambert, D. and C. Cooper, 2000. Issues in supply chain management. Industrial Marketing Management 29, 65-83.

Leap, T. and D. Grigsby, 1986. A conceptualization of bargaining power. Industrial and Labor Relations Review 39, 202-213.

Leaper, S., 1997. HACCP: A practical guide. Technical Manual 38. HACCP Working Group. Campeen food & drink research association. Gloucestershire, UK.

Lecraw, D., 1984. Bargaining power ownership, and profitability of transnational corporations in developing countries. Journal of International Business Studies 27 (5), 877-903.

Lee, J., W. Chen and Ch. Kao, 1998. Bargaining power and the trade-off between the ownership and control of international joint ventures in China. Journal of International Management 4, 353-385.

Lezema, J.H., 1989. Estudio comparativo de cuatro modalidades de aplicación de un insecticida y el uso de trampas de combate de las moscas de las frutas del género Anastrepha en Mango. Tesis Ingienería Agrícola Universidad de Costa Rica. 48 p.

Lindbloom, Ch., 1948. Bargaining power in price and wage determination. Quarterly Journal of Economics 62 (May), 396-417.

Luning, P.A., W.J Marcelis and W.M.F. Jongen, 2002. Food quality management: a techno-managerial approach. Wageningen Press, Wageningen.

Luning, P.A., W.J. Marcelis and W.M.F. Jongen, 2006. A techno-managerial approach in food quality management research. Trends in Food Science & Technology 17, 378-385.

Martinez, S.W., K. Smith and K. Zering, 1997. Vertical coordination and consumer welfare: the case of the pork industry. Food Consumption and Economics Division, Economic Research Service, U.S. Department of Agriculture. Agricultural Economic Report No. 753.

McLay, F. and T. Zwart, 1998. Factors affecting choice of cash sales versus forward marketing contracts. Agribusiness 14 (4), 299-309.

McQueen, J. and K. Miller, 1985. Target market selection of tourist: A comparison of approaches. Journal of Travel Research 24 (1), 2-6.

Menegay, M.R., B. Hutabarat and M. Siregar, 1993. An overview of the fresh vegetable subsector in Indonesia. Indonesian Agribusiness Development Project. ADP Working Paper No. 12. Jakarta.

Meijer, S., G. Zuniga-Arias and S. Sterrenburg, 2005. Learning experiences with the Mango Chain Game. In: Proceedings of the 9th international workshop on experimental learning in industrial management, May, Espoo, Finland.

Millet, D., 2003. The producers' organizations in the Costa Rican mango chain, Governance Structure and Performance. MSc. Thesis. Management Studies Group. Development Economics Group. Wageningen University. Wageningen.

Ministerio de Agricultura y Ganadería MAG., 2003. Censo Agrícola de la Región Pacífico Central. San José, MAG.

Mizrach, A., U. Flitsanov, Z. Schmilovitch and Y. Fuchs, 1999. Determination of mango physiological indices by mechanical wave analysis. Postharvest Biology and Technology 16, 179-186.

Moen, Ø., 1999. The relationship between firm size, competitive advantages and export performance revisited. International Small Business Journal 18, 53-72.

Molnár, P.J., 1995. A model for overall description of food quality. Food Quality and Preference 6, 185-190.

Molinero, L.M., 2003a. Meta-análisis. Asociación de la sociedad Española de Hipertensión. Liga Española para la lucha contra la Hipertensión Arterial. Marzo 2003.

Molinero, L.M., 2003b. Heterogeneidad entre los estudios incluidos en un meta- análisis. Asociación de la sociedad Española de Hipertensión. Liga Española para la lucha contra la Hipertensión Arterial. Diciembre

Montero, M.M. and M.M. Cerdas, 2000. Manejo poscosecha del mango para el Mercado fresco. 1st edition, San José. Centro de investigaciones Agronómicas, Laboratorio e Tecnologías poscosecha.

Moon, Ch. and A. Lado, 2000. MNC-Host Government Bargaining Power Relationship: A Critique and Extension Within the Resource-Based View. Journal of Management 26 (1), 85-117.

Moon, W., W.J., Florkowski, A.V.A., Resureccion, L.R. Beuchat and M.S. Chinnan, 1998. Consumer concerns about nutritional attributes in a transition economy. Food Policy 23 (5), 357-369.

Moore W.L., J.J. Louviere and R. Verna, 1999. Using conjoint analysis to help design product plataforms. Journal of Product Innovation Management 16 (1), 27 - 39.

Mora, J., J. Gamboa-Porras and R. Elizondo-Porras, 2002. Guía para el cultivo del mango (Mangifera indica) en Costa Rica. Ministerio de Agricultura y Ganaderia, Costa Rica.

Morisset, J., 1998. Unfair trade? The increase gap between world and domestic prices in commodity markets during the past 25 years. World Bank Economic Review 12 (3), 503-526.

Muradian, R. and W. Pelupessy, 2005. Governing the Coffee Chain: The Role of Voluntary Regulatory Systems. World Development 33 (12), 2029-2044.

Morgan, R.E. and C.S. Katsikeas, 1997. Export stimuli: Export intention compared with export activity. International Business Review 6 (5), 67-73.

Muthoo, A., 2000. A non-technical introduction to bargaining theory. World Development 1 (2), 145-166.

Muthoo, A., 2001. 'The economics of bargaining'. In: Knowledge for Sustainable Development: An Insight into the Encyclopedia of Life Support Systems, UNESCO and EOLSS: EOLSS Publishers Co. Ltd, 2002.

Muthoo, A., 2002. Bargaining Theory with Applications. Cambridge University Press.

Nash, J., 1950. The bargaining problem. Econometrica 18, 155-162.

National Advisory Committee on Microbiological Criteria for Foods (NACMCF) of the USA., 1998. Hazard analysis and critical control point principles and applications guidelines. Journal of Food Protection 61, 762-775.

Odulaja, A and F. Kiros, 1995. Modelling agricultural production of small-scale farmers in sub-Saharan Africa: A case study in western Kenya. Agricultural Economics 14, 85-91.

Ohyama, A., S. Braguinsky and K. Murphy, 2004. Entrepreneurial ability and market selection in an infant industry: evidence from Japanese cotton spinning industry. Review of Economic Dynamics 7, 354-381.

Ott, S.L., 1990. Supermarket Shoppers' Pesticide Concerns and Willingness to Purchase Certified Pesticide Residue-Free Fresh Produce. Agribusiness 6 (6), 593-602.

Ouchi, W.G., 1980. Markets, Bureaucracies, and Clans. Administrative Science Quarterly, 25 (1), 129-141.

Papadopolous, N. and J.E. Denis, 1988. Inventory, taxonomy and assessment of methods for international market selection. International Marketing Review 5, 38-51.

Papadopoulos, N., H. Chen and D.R. Thomas, 2002. Toward a tradeoff model for international market selection. International Business Review 11, 165-192.

Peacock, B.C., 1986. Postharvest handling of mangoes. Proceedings of the first Australian Research Workshop. CSIRO, Melbourne, 295 - 313.

Penrose, E.T., 1952. Biological analogies in the theory of the firm. The American Economic Review 42, 804-819.

Pil, F.K. and J.P. MacDuffie, 1996. The adoption of high-involvement work practices. Industrial Relations 35 (3), 423-455.

Pitchford, R., 1998. Moral hazard and limited liability: The real effect of contract bargaining. Economic Letters 61, 251-259.

Podolny, J., 1994. Market uncertainty and the social character of economic exchange. Administrative Science Quarterly 39, 458-483.

Pomareda, C., 2005. La agricultura en la economía y el desarrollo de Costa Rica, 1960-2004. In: Agricultura y desarrollo económico. Celebración de los cuarenta años de la publicación del libro Transforming Tradicional Agriculture de Theodore Schultz. Editores Grettel López y Reinaldo Herrera. Academia de Centroamérica, San José.

Ponte, S., 2002. The 'Latte Revolution'? Regulation, Markets and Consumption in the Global Coffee Chain. World Development 30 (7), 1099-1122.

Porter, M., 1990. The competitive advantage of nations. New York: Free Press.

Poulsen, C.S., H.J., Juhl, K, Kristensen, A.C. Bech and E. Engelund, 1996. Quality guidance and quality formation. Food Quality and Preferences 7, 127-135.

Prinsley, R.T. and G. Tucker, 1987. Chapter 1: Introduction. In: Mangoes: a Review. The Commonwealth Secretariat, London.

Rao, P.K., 2003. The economics of transaction cost. Theory, methods and applications. Palgrave Macmillan. London.

Randall, E. and D. Sanjur, 1981. Food preferences - their conceptualization and relationship to consumption. Ecology of food and nutrition 11, 151 - 161.

Raudenbush, S., and A. Bryk, 1985. Empirical bayes meta-analysis. Journal of Educational Statistics 10 (2), 75-98.

Roelofs, A.M.E., 2000. Structuring policy issues, testing a mapping technique with gaming simulation. Ph.D. thesis Katholieke Universiteit Brabant, The Netherlands.

Ray, D., 1998. Development Economics. Princeton University press.

Roets, M. and J.F. Kirsten, 2005. Commercialization of goat production in South Africa. Article in press. Small Ruminant Research.

Romano, G.S., E.D., Cittadini, B., Pugh and R. Schouten, 2006. Sweet cherry in the horticultural production chain. Stewart Postharvest Review 6 (2), 1-7.

Roumasset, J., 1995. The nature of the agricultural firm, Journal of Economic Behavior & Organization 26 (2), 161-177.

Rhim, H., T., Ho and U. Karkarkar, 2003. Production, Manufacturing and Logistics, Competitive location, production and market selection. European Journal of Operational research 149, 211-228.

Ruben, R, F. Saenz and G. Zúñiga-Arias, 2005. Contracts or rules quality surveillance in Costa Rican mango exports. In: G.J. Hofstede, H. Schepers, L. Spaans, J. Trienekens and A. Bielens (Eds.) Hide or Confide? The Dilemma of Transparency. 's Gravenhage: Reed Business, pp. 51-58.

Rubinstein, A., 1982. Perfect equilibrium in a bargaining model. Econometrica 50 (1), 97-109.

Rubinstein, A. and A. Wolinsky, 1985. Equilibrium in a market with sequential bargaining. Econometrica 53 (5), 113-1150.

SAMIC., 2002. Global meat market overview. South African Meat Industry Company.

Sáenz, F. and R. Ruben, 2004. Export contract for non-traditional products: Chayote from Costa Rica. Journal on Chain and Network Science 2 (4), 139-150.

Sáenz-Segura, F., 2006. Contract farming in Costa Rica: Opportunities for smallholders?, Phd Thesis, Wageningen University.

Schouten, R.E., G. Jongbloed, L.M.M. Trijskens and O. Van Kooten, 2004. Batch variability ans cultivar keeping quality of cucumbers. Postharvest Biology and Technology 32, 299-310.

Schouten, R.E., E.C. Omta, O. Van Kooten and L.M.M. Tijskens, 1997. Keeping quality of cucumber fruit predicted by the biological age. Postharvest Biology and Technology, 12, 175-181.

Secretaria Ejecutiva de Planificación Sectorial Agropecuaria (SEPSA), 2001. Boletín estadístico, sector agrícola costarricense. Ministerio de Agricultura y Ganadería, San José, Costa Rica.

Secretaria Ejecutiva de Planificación Sectorial Agropecuaria (SEPSA), 2002. Desempeño de la actividad del mango, 1996-2001. SEPSA, San José, Costa Rica.

Shan, R and P. Ward, 2003. Lean manufacturing: context, practice bundles, and performance. Journal of Operations Management 21, 129-149.

Shivery, G., 1997. Consumption risk, farm characteristics, and soil conservation adoption among low-income farmers in the Philippines. Agricultural Economics 17, 165-177.

Shewfelt, R.L., 1993. Measuring Quality and Maturity In: Postharvest Handling a Systems Approach, Academic Press Inc., San Diego, p. 99-124.

Shewfelt, R.L., 2006. Defining and meeting consumer requirements. Acta Horticulturae 712, 31-38.

Sijtsema, S., 2003. Your Health!? Transforming health perception into food product characteristics in consumer-oriented product design. Ph.D Thesis, Wageningen University.

Singh, S., 2002. Contracting out solutions: political economy of contract farming in the Indian Punjab. World Development 30 (9), 1621-1638.

Slichter, S., 1940. Impact of social security legislation upon mobility and enterprise. American Economic Review 30, 44-77.

Snell, S. and J. Dean, 1992. Integrated manufacturing and human resource management: a human resource perspective. Academy of Management Journal 35 (3), 467-504.

Sloof, M., L.M.M. Tijskens and E.C. Wilkinson, 1996. Concepts for modeling the quality of perishable products. Trends in Food Science & Technology 7, 165-171.

Sporleder, T.L., 1992. Managerial economics of vertically coordinated agricultural firms. American Journal of Agricultural Economy 74, 1226-1231.

Sporleder, T.L. and P.D. Goldsmith, 2001. Alternative Firm Strategies for Signaling Quality in the Food System. Canadian journal of Agricultural Economics 49 (4), 591-604.

Steenkamp, J-B.E.M., 1989. Product quality: An investigation into the cConcept and how it is perceived by consumers , Ph.D Thesis Wageningen University.

Sterrenburg, S. and G. Zuniga-Arias, 2004. The Costa Rica chain game. In: Hofstede *et al.* (Eds), Proceedings of the 8th international workshop on experimental learning in chain and networks, 24-27 May 2004, Wageningen.

Stinchcombe, A., 1985. Contracts as hierarchical documents. In: A.L. Stinchcombe and C. Heimer (Eds.) Organization theory and project management. Bergen: Norwegian University Press.

Sullivan, L.P., 1986. Quality function deployment. Quality Progress 19 (6), 39-50.

Talbot, J.M., 1997. The struggle for control of a commodity chain: Instant coffee in Latin America. Latin American Research Review 32 (1), 56-91.

Tenbrunsel, A., K. Wade-Benzoni, J. Moag and M. Bazerman, 1999. The negotiation matching process: relationships and partner selection. Organization Behavior and Human Decision Processes 80, (3), 252-283.

Teratanavat, R., V. Salin and N.H. Hooker, 2005. Recall event timing: measures of managerial performance in U.S. meat and poultry plants. Agribusiness, 23 (3), 351-373.

Terpstra, V. and C. Yu, 1988. Determinants of foreign investments of U.S. advertising agencies. Journal of International Business Studies 19 (2), 33-46.

Thompson, S.G. and S.J. Sharp, 1999. Explaining heterogeneity in meta-analysis: a comparison of methods. Statistics in Medicine 18, 2693-2708.

Tijskens L.M.M., P. Konopacki and M. Simcic, 2003. Biological variance, burden or benefit? Postharvest Biology and Technology 27 (1), 15-25.

Toornoos, J., 1991. Relations between the concept of distance and international industrial marketing. In: S. Paliwoda, New perspectives in international marketing. London: Routledge.

Trijp, J.C.M. and J.E.B.M. Steenkamp, 1998. Consumer-oriented new product development: principles and practice. In: W.M.F. Jongen and M.T.G. Meulenberg (Eds.) Innovation of Food Production Systems, Wageningen, Wageningen Pers.

UNTAD, 2000. Strategies for diversification and adding value to food exports: a value chain perspective. Study prepared by John Humphrey and Antje Oetero of the Institute of Development Studies, University of Sussex.

Van Boekel, M.A.J.S., 2005. 'Technological innovation in the food industry: product design'. In: Jongen, W.M.F. and Meulenberg, M.T.G. (eds) Innovation in agri-food systems, Wageningen Academic Publishers.

Weinberger, K. and J. Jütting, 2000. Women's participation in local organizations: conditions and constrains. World Development 29 (8), 1391-1404.

Weller, J., 1993. La generación de empleo e ingresos en las exportaciones no-tradicionales agrícolas: el caso de los pequeños productores de Centroamérica. In: PREALC, Maíz o Melón: Las Respuestas del Agro Centroamericano a los Cambios de la Política Económica. Panamá: ILO, 125-152.

Wenzler, I., 2003. 'The portfolio of 12 simulations within one company's transformational change initiative'. In: Arai, K (ed.) Social Contributions and Responsibilities of Simulation & Gaming, Chiba, Japan.

Williamson O., 1975a. Markets and Hierarchies: analysis of antitrust implications. New York: Free Press.

Williamson, O., 1975b. Transaction cost economics: the governance of contractual relations. Journal of Law and Economics 22, 233-261.

Williamson, O., 1981. The economics of organization: The transaction cost approach. American Journal of Sociology 87, 548-577.

Williamson, O., 1985. The Economic Institutions of Capitalism. The Free Press. New York.

Williamson, O., 1998. Transaction Cost Economics: How it works, where it is headed. De Economist 146, 23 - 58.

Williamson, O., 2002. The theory of the Firm as Governance Structure: From Choice to Contract. Journal of Economic Perspectives 16 (3),171-195.

Wilson, J., 1986. The political economy of contract farming. Review of Radical Political Economics 18 (4), 47-70.

White, R.E., J.N. Pearson and J.R. Wilson, 1999. JIT Manufacturing: a survey of implementation in small and large US manufacturers. Management Science 45 (1), 1-15.

WHO/FAO - World Health organization and Food and Agriculture Organization of the United Nations, 2005. Understanding the Codex Alimentarius. Editorial Production and Design Group. Publishing Management Service. FAO. Rome.

Yan, A. and B. Gray, 2001. Negotiating control and achieving performance in international joint ventures: A conceptual model. Journal of international Management 7, 295-315.

Zeithaml, V., 1988. Consumer Perceptions of Price, Quality and Value: A Means-End Model and Synthesis of Evidence, Journal of Marketing 52(July), 2-22.

Zingales, L., 1998. Corporate governance, In: J. Eatwell (Ed.) The New Palgrave, Macmillan, London.

Appendices

Appendix A. Characteristics of mango producers in different production areas in Costa Rica

	Esparza		Garabito		Orotina		Puntarenas		San Mateo	
	Mean	**σ**	**Mean**	**σ**	**Mean**	**σ**	**Mean**	**σ**	**Mean**	**σ**
Family size	5.2	3.03	4.33	2.16	5.25	1.16	4.7	1.83	3.17	1.72
Age household head (years)	50.8	12.56	49.67	14.06	47	14.47	49.5	14.71	36.5	13.28
Mango experience (years)	11	6.67	15.67	11.27	19.87	11.54	13.8	7.14	13.5	6.27
Mango area (ha)	4.40	3.66	4.94	5.09	19.36	22.57	9.71	12.02	27.05	37.69
Mango area/area total	0.34	0.18	0.45	0.29	0.90	0.15	0.64	0.47	0.77	0.38
Hired labour (000 colons)	1384	2107.3	573.3	930.6	1925.1	908.9	949.9	1154.1	1931.7	2635.7
Family labour (000 colons)	32	71.6	200	489.9	288	555	495.8	888	3600	5783
Ratio hired/family labour	43.3		2.9		6.7		1.9		0.5	
Input cost (000 colons)	1710	3516	15541	2291	971	1455	545	246	4950	888
Profits/loss (000 colons)	942	5865	1350	1511	-730	24175	3875	5933	2085	2438
Avg. annual price (colon/box)	2964	1066	2200	552.50	1838.4	1003.5	2495.7	807.2	2655.6	1639.7
Production (boxes/year)	1862	2090	1738	1847	4913	5561	3382	3804	5367	5602
Rejection in orchard (%)	7%	4.5	9.2%	20.1	3.8%	5.8	11.7%	11.9	5.3%	5.2
Yield (boxes/ha)	6.27	3.05	6.65	4.79	5.13	4.13	6.11	3.68	5.08	2.40

Appendix B. T-test for local and export market quality index differences

Quality index	t-value	Significance
Equal variance assumed	9.01	0.001***
Equal variance not assumed	9.02	0.001***

* significant at 10%, ** significant at 5%, *** significant at 1%.

Appendix C. Robustness test for the farm household, price and non-price characteristics

Variable	Coefficient	Significance	Robustness test		
Age	0.480	0.003***	✓	✓	✓
Experience	-0.198	0.231	✓	✓	✓
Mango area	0.619	0.001***	✓	✓	✓
Bargaining power	-0.650	0.001***	✓	✓	✓
Hired labour	-0.110	0.575	✓	✓	✓
Family labour	0.306	0.035**	✓	✓	✓
Input cost	-0.463	0.003***	✓	✓	✓
R^2	0.696		0.789	0.760	0.777
Rainy season price	-0.515	0.001***	✓	✓	✓
Dry season price	0.261	0.051*	✓	✓	✓
R^2	0.324		0.385	0.404	0.303
Important for reaching an agreement					
Traditional contract and friendship	0.326***		X	X	X
Amount control by the buyer	++				
Outcome characteristics	0.146		X	X	X
Relationship with buyer					
Trust and loyalty	-0.274**		X	X	X
Long-term relationships	0.460***		X	X	X
Informative and strict	++				
Determinants of outlet choice					
Justice and fairness	-0.909***		X	X	X
Friend and good reputation	0.634***		X	X	X
Better price, fast payment	-0.361***		X	X	X
Hard to bargain with	++				
Attributes influencing product management					
Technical package inputs and training	0.099		X	X	X
Friendship, information exchange	0.070		X	X	X
Hold-up	0.628***		X	X	X
Bargaining power attributes					
Risk taker	-0.426***		X	X	X
Commitment tactics	++				
External options	-0.523***		X	X	X
Wealth	++				
R^2	0.883		0.921	0.895	0.914

* significant at 10%, ** significant at 5%, *** significant at 1%.

Appendix D. Factor analysis results and weight of each factor for behavioural characteristics

Subject	Factors	Initial Eigenvalues	Variance explained
Factors influencing the conclusion of an agreement			62.76
	Traditional contract and friendship	4.81	29.85
	Amount of control by the buyer	1.54	19.30
	Outcome characteristics	1.19	13.61
Relationship with buyer			66.23
	Trust and loyalty	4.19	30.41
	Long-term relationships	1.35	22.55
	Informative and strict	1.09	13.28
Determinants of outlet choice			65.76
	Justice and fairness	4.69	27.71
	Friend and good reputation	1.47	13.62
	Better price, fast payment	1.27	13.05
	Hard to bargain with	1.12	11.38
Attributes influencing product management			65.08
	Technical package inputs and training	5.16	25.03
	Friendship, information exchange	1.65	24.54
	Hold-up	1.00	15.51
Bargaining power attributes			66.79
	Risk taker	2.26	20.58
	Commitment tactics	1.44	18.99
	External options	1.25	14.63
	Wealth	1.06	12.59

Appendix E. Construction of the main factors after performing the factor analysis

The five main factors were constructed as follows:

For the aspect *important for reaching an agreement*, we considered three factors:

1. traditional contract and friendship related to information about price, volume, payment, friendship and trust;
2. amount of buyer's control, consisting of visits from the buyer, buyer's willingness to give technical assistance, and transportation of the produce by the buyer;
3. outcome: characteristics related to quality, frequency of deliveries, and the technological package used.

For *relationship with buyers,* we selected three factors:

1. trust and loyalty, as expressed by the following statements: 'I trust my buyers', 'my buyer is loyal', 'my buyer trusts me', and 'I am loyal to my buyer';
2. long-term relationships related to having a friendship with the buyer, a long-standing relationship with the buyer, or a good overall relationship with the buyer;
3. informative and strict (referring to a buyer who provides the information the seller needs and who is strict in quality issues).

For the *determinants of outlet choice,* there are four factors:

1. justice and fairness characteristic of a buyer that gives credit, technical assistance, transport for the fruit, training, and information, but that is also concerned about the quality that the producer will deliver, and about being sure that the buyer will always buy from the producer;
2. friendship and a good reputation (including how long they have known the buyer and whether the buyer has a good reputation in the market);
3. better price and fast payment (consisting of a buyer that pays better than the market average, that pays promptly, and that never fails to pay);
4. hard to bargain with (linked to receiving fixed amounts of money for the total production and being able to bargain with the buyer).

For the aspect *attributes influencing product management* we used three factors:

1. technical package and inputs training (receiving specialised training on growing mangoes, buying new technology, getting technical assistance, using the inputs that the plantation needs, and producing for the export market);
2. friendship and information exchange (knowing the quantity of mangoes will be produced, knowing that the buyer will always buy, knowing that the buyer will always pay, and knowing how to do business with friends);
3. hold-up (related to the buyer's concern about what is being produced, about only producing for one buyer, and having a fixed price for the fruit).

For *bargaining power attributes*, we used four factors:

1. being a risk-taker (consisting of whether the producers perceived themselves as risk-takers, with no fear of losing a deal);
2. commitment tactics related to taking the necessary time to make a decision, to the main buyer being a good business partner, and to keeping one's word;
3. external options consisting of being patient, or having many buyers;
4. wealth (related to producers that have economic resources and are able to bargain prices with the buyer).

Appendix F. Construction of indexes

Index	Variables	Average	Standard deviation
Quality	Brix/pH	2.6504	0.8151
	Brix	10.0340	3.8070
	pH	3.7222	0.4131
General Management Intensity		3.8235	1.2577
	Planning	3.6471	1.6592
	Organisation	3.6667	1.7050
	Control and monitoring	3.6667	1.8619
	Marketing	4.1373	1.0958
	Finance	4.0000	1.5362
Quality Management		4.0039	0.8212
	Temperature	4.1961	1.1836
	Selection	4.3529	0.9965
	Packaging	4.4510	1.0260
	Activity registry	3.2157	1.7008
	Storage	3.8039	1.5364
Access to Information		4.2843	0.6318
	Produce price	4.8039	0.4481
	Market place	4.3137	0.9896
	Quality and quantity of inputs	4.0196	1.3189
	Pests and diseases	3.8824	1.4785
	Quality standards	4.2549	1.0741
	Reputation	4.4314	0.9645
Operation Index		4.6422	0.6389
	Adequate fertilisation	4.3137	1.2408
	Adequate harvesting	4.6275	0.8936
	Adequate selection	4.8431	0.4182
	Adequate transport	4.7843	0.8322
Technological variation index		4.0280	1.0134
	Genetic homogeneity	3.1176	1.7280
	Quality sampling	4.0196	1.4351
	Agro-environmental conditions	4.1176	1.2607
	Input use	4.3137	1.1746
	Harvesting time	4.4800	1.0150
Socio-economic Index		4.4980	0.5357
	Agreement for buying and selling	4.3333	1.0893
	Defined price before business	4.0784	1.4946
	Market knowledge	4.7451	0.4835
	Trust among partners	4.5294	1.0070
	Long relationships	4.8039	0.5664

Appendix G. Variability of the main variables for the analysis

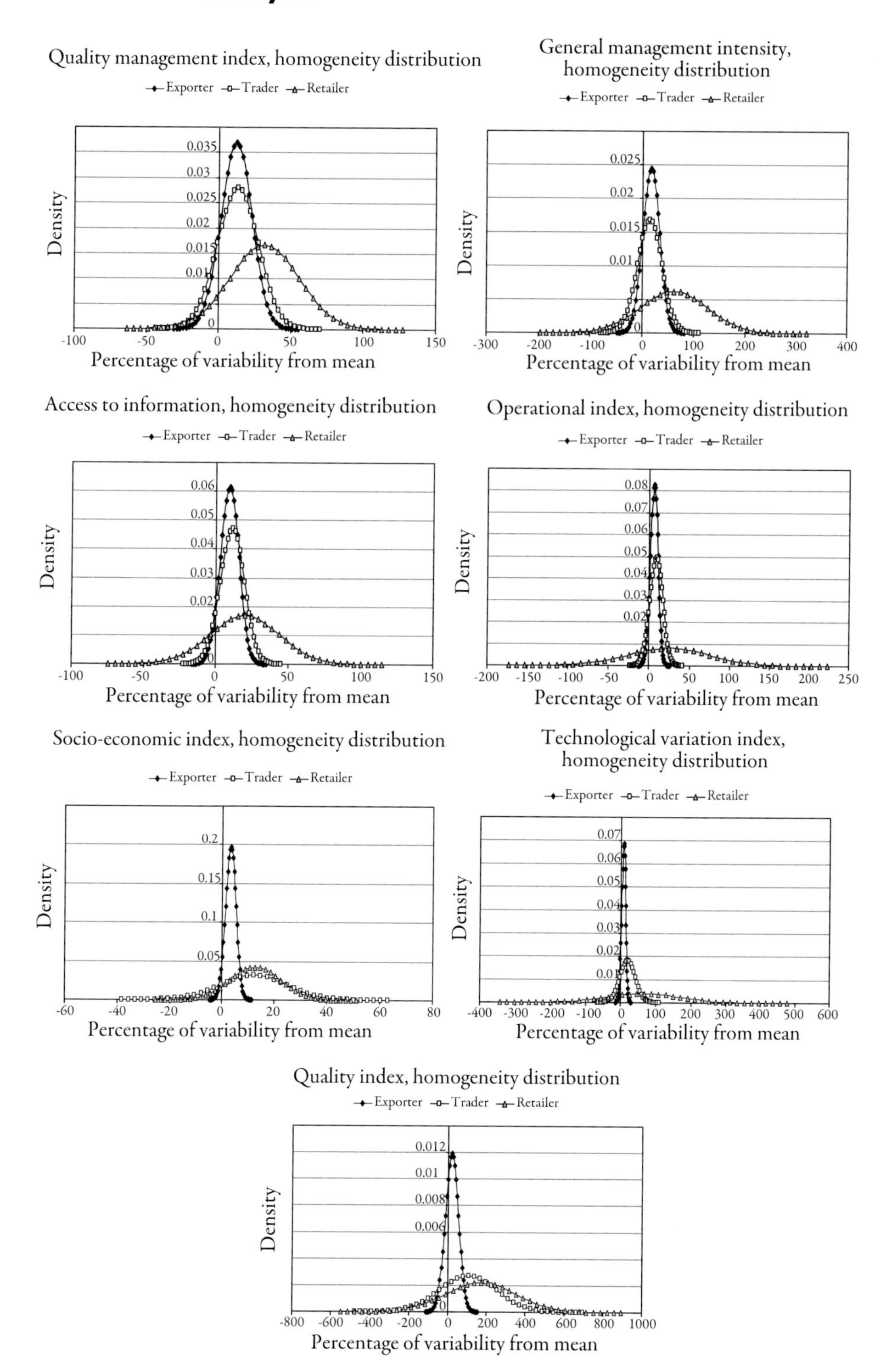

Appendix H. T-test for the mean differences between the main variables in the analysis and the different actors

	Exporter		Trader		Retailer	
	T-value	P-value	T-value	P-value	T-value	P-value
Quality	-6.616	0.0001***	2.994	0.0042***	1.836	0.0808*
GMI	1.485	0.1472	0.955	0.3412	-1.788	0.0921*
QM	1.34	0.1863	0.555	0.5823	-1.458	0.1650
AI	0.592	0.5575	-0.622	0.5387	0.056	0.9556
OI	0.959	0.3447	0.167	0.8691	-0.753	0.4620
TV	1.776	0.0811*	0.045	0.9649	-1.003	0.3348
SE	2.628	0.0106**	-0.387	0.7023	-1.237	0.2391

*** $p < 0.01$, ** $p < 0.05$, * $p < 0.1$.

Summary

Supply chain analysis is of vital importance for understanding business relationships and strategic decisions of different actors involved in the production, distribution and consumption of commodities. The mango supply chain in Costa Rica provides a good illustration for analysing supply chain integration and agency coordination. We analysed several key issues to understand the relationships and transactions among actors, giving special attention to the quality performance of mangoes, the variability of management and quality, market selection criteria, bargaining power determinants and revenue distribution.

This research was conducted in Costa Rica, a small country located in Central America, with an open economy based on the export of fresh products, computer chips and tourism. The field work was carried out in the Pacific region of the country, and included two and half years of data collection from three surveys, two mango laboratory analyses and five game simulation games with real mango producers.

The methodological approach of this research is based on an integrated framework linking criteria from natural and social science. We collected mango samples and analysed their intrinsic and extrinsic characteristics, and related the outcomes to different socio-economic characteristics of farm households, their production system, the type of delivery agreements, and the product management practices.

We also used a game simulation approach where the innovation is to use it as a learning tool for research on exchange decisions. Producers were invited to play a role in the simulation and hence provided information regarding their perceived bargaining power and revenue distribution. The advantage of this tool is that real producers are participating and perform naturally according to how they do business in reality. It allows the researcher and producers to interact and learn more about the performance and organisation of the supply chain to which they belong; producers can also observe how the different business strategies operate in a simulation lasting a couple of hours.

The results of the present study are divided into four issues: market choice, economic incentives for quality, bargaining power and revenue distribution, and quality and management variability.

Market outlet choice. Different actors are involved in the mango supply chain in Costa Rica conducting transactions in both the export and local markets Producers face a strategic choice between (a) market outlets devoted to exports where quality attributes such as size, sugar content, and no external or internal damage are key determinants for transactions and business relationships, and (b) local markets (composed of wholesalers, retailers, and middlemen among others) where different qualities and delivery modes can be accommodated.

We argue that market selection is dependent on farm-household characteristics, the structure of the production system, price attributes and the market context. The selection of a certain type of outlet is also related to specific contract configurations (i.e. procedures for quality control, payment mode, type of agreement, volume, rejection rate, etc.). For this purpose, we have disentangled factors influencing market differentiation between the export and local market in the Costa Rican mango sector, and we could identify the main determinants for differences in outlet choice.

Based on a field survey regarding outlet choice decisions amongst a sample of 94 farmers in the major mango-producing region (Central Pacific area) of Costa Rica, we have used a Tobit model and ANOVA analysis to assess the determinants of market outlet choice. We used structural, institutional and behavioural factors that typically determine farmers' choices for a specific market channel orientation. We find that for the export market vertical integration is already fairly advanced: producers deliver on the demands of the buyer and face higher rejection rates, but in compensation they receive access to stable market outlets, input and (subsidized) credit, and benefit from lower transport and delivery costs. In the local market, the producers' experience and their historical knowledge and relationships with the market are of key importance for finding a suitable market outlet. Although we relied on rather standard analytical procedure, for the particular case of mango we found that some farm household characteristics (notably farmers' experience and risk-averse attitudes) explain export market outlet selection, contrary to the findings in some other studies in the literature reviewed.

Economic incentives for quality We developed an integrated methodology for identifying effective economic incentives to enhance quality performance amongst mango producers in Costa Rica. Therefore, we analysed the relationship between intrinsic quality characteristics measured at the field level and the socio-economic characteristics of the mango producers in the Central Pacific area of Costa Rica. Data are derived from a representative sample of 35 mango producers where the quality of five selected mango fruits has been assessed. A mango quality index for specified export market outlets is constructed and subsequently related to a set of socio-economic characteristics at farm and household level.

We used categorical regression methods to identify the relationships between farm-household characteristics, production system features, marketing relationships and quality attributes measured by consumers' quality perceptions and preferences. Subsequently, we designed a model where subjective quality is related to technical and institutional aspects concerning the organisation of mango delivery systems.

The general attributes of the quality index – mainly the dimensions of ripeness, appearance and variability – appear to be strongly related to farm-household characteristics, including the producers' age and experience, input use intensity and family labour availability. In addition, preferences for certain contractual regimes and marketing arrangements gives rise to a differentiation in quality performance. Long-term relationships and non-price attributes appear to be key factors in mango quality improvement.

Bargaining power. Transactions between producers, traders, retailers and consumers in the mango supply chain are governed by contractual arrangements concerning outlet choice, price, volume, quality and delivery frequency. We developed a stylised gaming simulation for the appraisal of different sets of delivery transactions among mango supply chain agents. The gaming simulation design closely mimics the negotiation conditions in the Costa Rican mango supply chain. Agency roles were defined for all participants in the chain and the attributes of all of the transactions have been recorded. The gaming simulation was played five times with different groups of Costa Rican mango producers, resulting in a data set of 82 transactions and 43 bargaining power positions. Bargaining power was assessed on a 10-point Likert scale and revenue distribution was measured in terms of value added. The results show that bargaining power is dependent on the negotiation skills, wealth and good partnerships with the negotiators, but it is independent of market imperfections. Revenue distribution is related to the perceived bargaining power of the trading partners, risk perceptions and the length of the contract. We conclude that gaming simulation can be considered as a promising experimental research method. Transparency and agency cooperation play a key role in improving the efficiency of mango supply chains in Costa Rica as seen from the producers' perspective. Initiatives to improve the bargaining power of producers might focus especially on better bargaining skills and relations in trade rather than on directly solving market imperfections.

Variability in quality and management. Quality is a key aspect for evaluating the performance of commodity chains, but quality management is a highly complex procedure. Quality refers to consumer perception regarding a certain product as well as the objective relationship between intrinsic and extrinsic attributes of the product. Improved cropping operations and general management activities may enhance the quality and/or reduce the variability of the produce. In addition to these technological measures, timely access to information and improved quality management operations can be helpful to reduce human-related variability.

In this research we present an exploratory framework for disentangling the interactions between different managerial activities that have an effect on the variability of quality. For this purpose, we use several data dispersion statistics to understand the impact of technological and socio-economic factors on heterogeneity in quality performance at different stages of the supply chain. We have conducted a field survey amongst 51 stakeholders involved in the mango supply chain from Costa Rica and collected information regarding production technologies, agro-ecological conditions, management intensity, quality control, contracting practices and marketing operations. We also collected 10 mangoes randomly from each actor to analyse the variability in intrinsic quality attributes (defined as the ratio between Brix and pH).

We find that quality variability is influenced both by technological and socio-economic variation. In the mango supply chain in Costa Rica, the management differences amongst actors vary depending on closeness to the consumer market. Actors closer to the market tend to maintain higher variability in their management practices in order to be able to respond better to market challenges.

The thesis used an interdisciplinary research framework to assess options for improving quality and reducing the variability in quality through adjustment in management and marketing practises. It shows that the local market is highly complementary to the export market and that personal producers' characteristics and wealth are key factors that influence the bargaining power of mango producers in Costa Rica.

Curriculum Vitae

Guillermo Zúñiga-Arias was born in San José Costa Rica the 2nd of June, 1974. He studied at the *Universidad Nacional* and received a Bachelor's degree in Social an Economical Planning (2000). He went on to obtain an M.B.A. with emphasis on Marketing (2002). In January 2003 he began his PhD studies in the Development Economics Group of the Department of Social Sciences at Wageningen University and Research Center, the Netherlands.

Guillermo has been working with formal and voluntary organisations, promoting organizations and project management. He has been working since 2000 at the Universidad Nacional with the *Escuela de Planificación y Promoción Social* and the International Centre for Political Economy and Sustainable Development (CINPE). At CINPE he worked as assistant and researcher in supply chain analysis studies. He is currently involved in teaching and research at the *Universidad Nacional.*

Printed in the United States
by Baker & Taylor Publisher Services